BASICS OF ELECTRON OPTICS

BASICS OF ELECTRON OPTICS

DAVID A. DE WOLF
Bradley Department of Electrical Engineering
Virginia Polytechnic Institute and State University
Blacksburg, Viginia

A Wiley Interscience Publication
JOHN WILEY & SONS, INC.
New York / Chichester / Brisbane / Toronto / Singapore

Library of Congress Cataloging in Publication Data:

De Wolf, David A.
Basics of electron optics / David A. de Wolf.
p. cm.
"A Wiley-interscience publication."
Includes bibliographical references (p.
ISBN 0-471-52457-3
1. Electron optics. I. Title.
QC793.5.E62D48 1990
537.5'--dc20 90-12437
CIP

Printed in the United States of America

10 9 8 7 6 5 4 3 2 1

CONTENTS

LIST OF SYMBOLS

$\vec{A}$	vector magnetic potential
$\vec{B}(\vec{r}, \omega)$	the electric flux density at $\vec{r}$ as a function of frequency
C_{fa}	spherical-aberration coefficient in focal plane
C_{ia}	spherical-aberration coefficient in image plane
C_1, C_3	coefficients of optimal ρ_B, ρ_s design
$\vec{D}(\vec{r}, \omega)$	the magnetic flux density at $\vec{r}$ as a function of frequency
$\vec{E}(\vec{r}, \omega)$	electric field vector at $\vec{r}$ as function of frequency
$\mathscr{E}_k$	relativistic kinetic energy $[=m_0c^2(\gamma-1)]$
f	image focal distance
f_0	object focal distance
$f_N(\cdots)$	phase space density of specified arguments in $6N-D$ space
$F(\cdots)$	4-D phase-space density of transverse coordinates
$\vec{F}$	Newtonian force vector
$\vec{H}(\vec{r}, \omega)$	magnetic field vector at $\vec{r}$ as a function of frequency
$\mathscr{H}(s)$	the Hamiltonian as function of trajectory coordinate s
I	current in A
$\vec{J}$	current density vector in A/m^2
k	wavenumber $(=2\pi/\lambda)$
k_B	Boltzmann's constant
K	perveance $(=\beta I/V^{3/2})$
$\vec{K}$	magnetic curvature $(=e\vec{B}/m\dot{z})$
L	image distance to P principal plane

L_0	object distance to P_0 principal plane
$\mathscr{L}(\vec{\rho}, \vec{\rho}', z)$	the Lagrangian for two-dimensional space, one-dimensional "time" coordinate
M	magnification
$n(\vec{r}, \omega)$	the refractive index at $\vec{r}$ as a function of frequency
$\vec{p}$	momentum vector $(=m\dot{\vec{r}})$
$\vec{p}_T$	transverse momentum vector
p_z	longitudinal momentum
$\vec{P}$	momentum measure ($=\vec{p}/\sqrt{2m_0}$, measured in $\sqrt{\text{eV}}$)
P	image principal plane intersection with lens axis
P_0	object principal plane intersection with lens axis
q, Q	charge (in C)
$\dot{\vec{r}}$	velocity (with components $\dot{x}$, $\dot{y}$, $\dot{z}$)
$\vec{r}_s$	surface-location coordinate
R	difference coordinate, distance, radius
R_{PQ}	distance between points P and Q
$R(z)$	the Picht reduced coordinate $(=\rho(z)[\Phi_{ao}(z)]^{1/4})$
$\mathscr{R}(\rho, \rho'; \Phi_0)$	term in trajectory equation for 3rd-order aberrations
v	velocity (alternative notation)
z	the cartesian coordinate chosen as reference direction
s	the trajectory coordinate $(=vt)$
s	lens strength parameter
S	lens strength parameter for Picht-coordinate lens formula
$S(\vec{r})$	eikonal function of geometrical optics
T	kinetic energy $(\frac{1}{2}mv^2)$
$T(z)$	logarithmic axial potential derivative $[=\Phi_0'(z)/\Phi_0(z)]$
U	potential energy $(q\Phi)$
U_k	equivalent kinetic energy $(=p^2/2m)$
V	potential voltage difference (usually as maintained on a metal plate)
β	the relativistic velocity v/c
γ	the relativistic parameter $(1-v^2/c^2)^{-1/2}$
Δ	the Laplace operator in three dimensions $(\partial^2/\partial x^2 + \partial^2/\partial y^2 + \partial^2/\partial z^2)$
ε	dielectric permittivity
θ	angle (polar)
λ	wavelength
μ	magnetic permeability
$\vec{\rho}(z)$	the transverse-direction coordinate vector at z
$\vec{\rho}'(z)$	the direction tangent of the trajectory at z
ρ_B	beam radius at end of main lens
ρ_s	screen-spot radius

ϕ	angle (azimuthal)
$\Phi(\vec{r})$	the electrostatic potential at $\vec{r}$
$\Phi_0(z)$	abbreviation for the axial potential $\Phi(\vec{0}, z)$
$\Phi_a(s)$	the accelerating potential at s $(=mv^2/2e$ for an electron)
Φ_i	constant potential on object side of lens
Φ_f	constant potential on image side of lens
Φ_m	scalar magnetic potential
Ψ	magnetic flux
ω	angular frequency in rads/s, or charge density in C/m^3
$\omega_s(\vec{r}_s)$	surface-charge density at a surface location $\vec{r}_s$
Ω	rotation frequency $(=eB_z/m)$
$\vec{\nabla}$	the gradient operator (del, nabla)

PREFACE

Electron optics is a subject to which I was first introduced when, at RCA Laboratories at the David Sarnoff Research Center in Princeton, New Jersey, I began to work on television kinescope electron-gun design by virtue of having some familiarity with wave optics. Consequently, I tried to learn as much of the subject as I could from existing textbooks. In 1975, the main ones available to me were a relatively old text by Zworykin et al. (1954) (with relevant chapters written by Ramberg), equally ancient texts by Cosslett (1946), Maloff and Epstein (1938), Pierce (1954), Klemperer (1953), an authoritative text in terms of an elegant Hamiltonian formalism by Sturrock (1955), and finally, one more-or-less "normal", more modern text, a revised version (1972) of Grivet's text. There were also several helpful texts on electron physics (Klemperer and Barnett 1975), phase-space approaches (Lichtenberg 1969, Lawson 1977), matrix formalism (Dahl 1973), space-charge optics (Kirstein et al. 1967, Nagy and Szilagyi 1974). Since then, several new texts or revisions of older texts have appeared (El-Kareh and El-Kareh 1970, Wollnik 1987, Szilagyi 1988, Hawkes and Kasper 1989). Unfortunately, most texts available in 1975 are now out-of-print or very highly priced. The above list is by no means complete, and there is no intention of slight towards any author(s) whose works I have omitted in this very brief summary.

However, I was only trying to summarize for myself what was available in 1975 when I set out to learn as much as I could about electron optics.

The use of an electronic computer was relatively recent, and almost none of the available texts made reference to computer simulations, whereas the applications I was involved with required use of then-existing main-frame computer programs incorporating the main ideas of electron optics. Therefore, I needed more mathematical detail than could ordinarily be found in the existing texts.

I found it frustrating to hunt back and forth from text to text for bits and pieces of what I needed. Moreover, although I had more than ten years of industrial experience at Ph.D. level, I found that I could not follow the line of reasoning in any of the text books without a great deal of hard (and, in my opinion, unnecessary) work. I was not alone; younger engineers and scientists who came after me had similar complaints. Most, indeed just about all, of the texts with the exception of Dahl's and Sturrock's, seem to be strongly oriented towards applications. In their effort to get to the applications, the authors leave much work essential to understanding the underlying physics and mathematics to the reader. A reader such as myself, however, is also anxious to be able to apply the material, and is therefore apt to be frustrated when the going gets rough and the text is of little help. The following example from the excellent text by Grivet (1972) is typical, and I hasten to add that there is no intention to find fault with it, as Grivet's intention is quite different from mine. In discussing the reduction of six to five aberration coefficients for axisymmetric lenses, Grivet in Section 7.2.1 refers to a relatively special, but not excessively difficult to understand symmetry property as a "theorem of Malus and Dupin". It took me days to decipher this seemingly simple statement, and it only made sense to me when I discovered after much reading on the subject that this property follows from the fact that the optical eikonal function is not a function of velocities! The eikonal function is central to geometrical optics, and electron optics is analogous to geometrical optics. For the reader who desires an outline of theory (and does not much care about much detail in deriving theory from the fundamentals) this does not matter. But for readers such as myself, it does. I therefore resolved to write a text that would spare the reader frustrations similar to mine, yet would avoid the rather specialized worlds of Dahl's and Sturrock's texts. The present text is the outcome.

The text is aimed at a fairly wide range of readers. In the first place, it is aimed at scientists and engineers who find they must learn something about electron optics in connection with design work using cathode-ray tubes (CRTs), electron microscopes, spectrometers, and other devices involving motions of beams of charged particles. I believe they will find this text more accommodating than others. These users will in all likelihood not have had a course in electron optics because there are not many currently being offered at American universities. At present (the late 1980s), design of electron optics in CRTs is of heightened interest for two reasons: the imminent introduction of high-definition television which requires screen displays of finer resolution than are presently available, and the proliferation of the CRT monitor for table-top, lap-top, and portable personal computers. For all these reasons, we find there is a new interest in electron optics, and there are no texts available at the level of this one that are based upon this need. Because the text is an outgrowth of a one-semester graduate course given in the Bradley Department of Electrical Engineering at VPI&SU, I have organized the material so that it will also serve as an early graduate text in either an applied physics or an electrical engineering curriculum. I hope it will stimulate the initiation of such a course in more universities. As stated above, I mainly had in mind the graduate engineer, physicist, or other initiate into the industrial environment who is confronted with the need to learn electron optics. Such readers will in all likelihood have sufficient background in the underlying electrostatics and physics to be able to work themselves into the subject from the present text. Then there are the practising engineers and scientists who need to brush up on or acquaint themselves with the underlying principles. The conciseness of this text and the detail with which the material is treated should appeal particularly to this class of readers.

The present text is aimed somewhat more at electrostatic rather than at magnetic lenses. This reflects the experience of the author, but the emphasis is upon basics and principles, and much of what seems peculiar to electrostatic lenses also applies to magnetic lenses. A final section on kinescope electron-gun design is perhaps more specialized than the average reader will want, but the material does not (yet) seem to be gathered elsewhere. Uninterested readers can skip this and several other sections without losing the trail.

Excellent texts on numerical (computer) calculations in electron optics exist, e.g. Hockney and Eastwood (1981), McAllister et al. (1985), and Steele (1987). I wished to avoid the Scylla of complete avoidance and the Charybdis of numerical detail, and so I chose to work out whatever mathematical detail was needed short of encoding it in a computer language. Numerous codes exist elsewhere that allow simulation of particle trajectories from kinematic force equations for given forces or potentials, and both trajectory equations and potentials are described in detail in the present text. Today's students and workers should not encounter much difficulty either in obtaining or working out computer programs for trajectories once the necessary equations are given.

Finally, a primary goal of mine was to be as clear as possible, and I can only hope that I will have succeeded. Good texts need to grow organically through feedback, and I expect and hope to receive comments that will enable me to improve on this one in future editions (should the need for such arise).

It remains only to express my gratitude to those who helped me learn electron optics. Among these are some of my former colleagues at what was once RCA's David Sarnoff Research Center in Princeton, NJ, and this gratitude extends warmly to the authors of those books that inspired me to add this one to the list of texts on electron optics.

D. A. DE WOLF

Blacksburg, March 1990

REFERENCES

Cosslett, V. E. (1946) *Introduction to Electron Optics*, 2nd edn., Oxford University Press, Oxford.

Dahl, P. (1973) *Introduction to Electron and Ion Optics*, Academic Press, New York.

El-Kareh, A. B. and El-Kareh, J. C. J. (1970) *Electron Beams, Lenses, and Optics*, Academic Press, Orlando, FL.

Grivet, P. (1972) *Electron Optics*, 2nd edn., Pergamon Press, Oxford.

Hawkes, P. W. and E. Kasper (1989) *Principles of Electron Optics*, Academic Press, Orlando, FL.

Hockney, R. W. and J. W. Eastwood (1981) *Computer Simulation Using Particles*, McGraw-Hill, New York.

Klemperer, O. (1953) *Electron Optics*, Cambridge University Press, Cambridge.

Klemperer, O. and M.E. Barnett (1970) *Electron Optics*, Cambridge University Press, Cambridge.

Kirstein, P. T., G. S. Kino and W. E. Waters (1967) *Space-Charge Flow*, McGraw-Hill, New York.

Lawson, J. D. (1977) *The Physics of Charged-Particle Beams*, Clarendon Press, Oxford.

Lichtenberg, A. J. (1969) *Phase-Space Dynamics of Particles*, John Wiley, New York.

Maloff, I. G. and D. W. Epstein (1938) *Electron Optics in Television*, McGraw-Hill, New York.

McAllister, D., J. R. Smith and N. J. Diserans (1985) *Computer Modelling in Electrostatics*, Research Studies Press, John Wiley, New York.

Nagy, G. A. and M. Szilagyi (1974) *Introduction to the Theory of Space-Charge Optics*, MacMillan, London.

Pierce, J. R. (1954) *Theory and Design of Electron Beams*, 2nd edn., Van Nostrand, New York.

Steele, C. W. (1987) *Numerical Computation of Electric and Magnetic Fields*, Van Nostrand Reinhold, New York.

Sturrock, P. A. (1955) *Static and Dynamic Electron Optics*, Cambridge University, Press, Cambridge.

Szilagyi, M. (1988) *Electron and Ion Optics*, Plenum Press, New York.

Wollnik, H. (1987) *Optics of Charged Particles*, Academic Press, Orlando, FL.

Zworykin, V. K., G. A. Morton, E. G. Ramberg et al. (1954) *Electron Optics and the Electron Microscope*, John Wiley, New York.

BASICS OF ELECTRON OPTICS

CHAPTER 1

INTRODUCTION

As stated in the Preface, the reader who wishes to delve into electron optics finds that there are only a few recent texts on the subject. The texts by Wollnik (1987) and by Szilagyi (1988) are available at the time of writing (1989–1990); whatever other texts are available appear to be somewhat specialized. The present text differs from these in that it treats mostly the essentials, and it attempts to give more help to the uninitiated reader than existing texts. As a result, it is not exhaustive in treating various kinds of electrostatic and magnetostatic lenses. Nothing is said specifically about the electron microscope, which is treated at a relatively elementary level in a text by Hall (1953). Grivet (1972) treats a number of applications, but his text is out-of-print and significant progress has been made in the electron optics of cathode-ray tubes (CRTs) since 1972. The practical text by Moss (1968) is of historical interest in this respect, but the advent of computer simulation has, perhaps, dated many of the numerical methods described therein. Nevertheless, this text is highly recommended to the reader as an adjunct to Chapter 11 of the present text as a supplement to the empirical laws distilled from more modern computer simulations.

The present text consists of nine chapters dealing with various aspects of electron optics, and the final chapter deals specifically with application to CRTs. A survey of the contents is given in this introductory chapter.

Chapter 2 establishes the close relationship between electron optics and classical geometrical optics of waves. The key point is that the ray equation of geometrical optics, which describes the spatial structure of the "rays" (i.e., the orthogonal trajectories to the wave fronts) of geometrical optics, is identical to that which describes the electron trajectories in static electromagnetic fields. The role of the refractive index of optics is mirrored exactly by the square root of the accelerating potential of the charged particle in beam optics. These two quantities vary differently in space; one is a material property and the other is an imposed field in space. Therefore, there are also some differences between beam and wave optics. The differences are highlighted.

The first issue of interest in beam optics is the spatial structure of a beam of charged particles in an electromagnetic field. Because the transition time of an electron across a CRT is usually short compared to the times in which the circuitry needs to modulate the electromagnetic fields that guide the beam, it usually suffices to deal with *static* fields. The beam is essentially static during the transition time of an electron across the devices of interest. Thus it is the *spatial* structure of the beam, consisting of stationary trajectories of individual particles, that is our major interest. Chapter 3 deals with the equations of motion that establish these trajectories for given static fields $\vec{E}$ and $\vec{B}$. We show that as long as accelerating voltages are of the order of 30 kV or less, it suffices to use nonrelativistic mechanics of particle motion if errors less than 1.5% or so can be tolerated. The question of how tolerable an error can be is certainly a function of the application, but present-day design does not really require greater accuracy in most applications to shaping beam spots on a CRT screen, because many other factors give rise to a much greater variation with position on a screen and with respect to manufacturing tolerances. An alternative to the ray-tangent variables, is the concept of particle momentum. This is introduced in Section 3.2 in preparation for the phase-space description of beams in Chapter 9. The relativistic description is given in Section 3.3. The derived equations of motion are not fully invariant with respect to the relativistic formulation because there is no need for a description of frames other than the one in which the fields are at rest. Two specific geometries of major importance in electron optics are that in which there is cylindrical symmetry, and that in which there is planar symmetry (no dependence of fields upon one Cartesian coordinate), and mirror symmetry with respect to a plane through the principal and the inert-coordinate axis. These geometries allow simple expansions of the electrostatic potential around the axial value. A brief discussion is given of numerical integration of the equations of motion.

In Chapter 4, we summarize the main methods of calculating the electromagnetic fields that are needed in the equations of motion. Various subsections take the reader through the basics of simple analytical methods, separation-of-variable techniques, the method of images, and conformal transformations (specifically, the Schwarz-Christoffel transformation for which it seems difficult to find adequate exposition in the mathematical literature!). The section ends with a brief survey of numerical methods. Chief among these, for electrostatics, are finite-mesh relaxation methods (for situations with well-defined closed-surface boundary values of potentials or fields) and integral-equation methods (for "open" geometries). Finite-element methods are similar in principle to finite-mesh relaxation methods and they are not discussed here. An integral-equation method, novel to our knowledge, is proposed for calculating the magnetic field for given current sources in the presence of locally uniform magnetic materials.

The preceeding chapters lay the framework for calculating charged-particle trajectories in static electromagnetic fields. In Chapter 5, we outline the classical first-order results of electron optics. By "first-order", we mean Gaussian or paraxial optics, obtained from the equations of motion by approximations valid for small and slow deviations from the main propagation axis. All of the classical formulas for either thin or weak lenses follow from the linear paraxial equations for electrostatic fields. The Picht formulation of equations for the reduced transverse ray coordinate $R = \rho\phi^{1/4}$ is discussed and compared to the classical one for ρ. Magnetostatic equivalents are discussed in Sections 5.3–5.5.

The actual lens formulas are discussed in Chapter 6. We derive the locations of principal planes and foci, as well as the thin-lens formulas for cylindrically symmetric electrostatic lenses in Section 6.1. The solutions for thicker lenses are discussed subsequently. Basic properties such as the reversal of principal planes and the general focussing properties of accelerating lenses are derived from the previously found properties. Finally, some elementary analogies for cylindrically symmetric magnetic lenses are formulated.

Chapter 7 allows us to treat several classes of lenses of importance to the electron guns in CRTs. Foremost among these are cylinder and aperture lenses. Exact solutions for the interior fields of cylinder lenses can be given only as two-fold integrals, but a fortuitous approximation allows us to write the solutions not too far from the axis in terms of a closed-form expression without sacrifice of great accuracy. The fields due to one electrostatic aperture lens with asymptotically uniform fields on either side can be given by a closed-form expression. An approximate analysis in terms of the asymptotic fields and plate voltage is discussed. Compound aperture lenses require an approximate linear combination of the one-aperture formulas, which is useful for description of electron gun aperture lenses (see Chapter 11). The subject of strong focussing is broached in Section 7.4 in which the class of quadrupole lenses in particular is singled out, as it arises naturally from the analysis.

Aberrations, i.e., deviations from paraxial optics, is a somewhat outmoded subject because the advent of numerical simulations makes it possible to obtain accurate trajectories even at appreciable off-axis distances. We follow a straightforward expansion through third-order in off-axis distances and ray tangents of the trajectory equations to find spherical aberrations in the focal and in the image planes. In Section 8.2, we follow an approach similar to Grivet (1972) (however, we do not need the Hamilton characteristic function) to obtain some general expression for aberrations of axisymmetric systems. Spherical aberration, coma, curvature, astigmatism, and chromatic aberrations are discussed briefly.

It is useful to introduce the concept of phase space in electron optics because, among other things, the transition of particles through lenses can be described by relatively simple coordinate transformations in phase space (at least in the paraxial approximation). In Chapter 9 we outline how the general $6N$-dimensional phase space of Liouville's theorem for an N-particle beam can be reduced to a four-dimensional one for the individual particles. Integration over both transverse momentum coordinates of the 4-D density in that phase space gives rise to the current density. Section 9.4 treats several elementary transformations that describe drifts and momentum changes by lenses in the paraxial approximation. The concepts of acceptance and emittance are touched upon briefly.

Chapter 10 deals with the problem of the mutual repulsion of electrons in a beam due to the Coulomb forces between them: the *space charge* problem. Space charge requires the numerical analyst to iterate the equations of motion, constantly laying down new approximations of the space charge from the trajectories of the previous iteration. The process will converge in general, but it may require many iterations (expensive!), and, in the case of long narrow beams, may require very many trajectories in order to avoid the apparent (but not real) singularity at locations where trajectories come close to each other. Using the well-known example of uniform laminar beams (in Sections 10.1 and 10.2), we prepare for a general numerical approximation (de Wolf 1982) in Section 10.3. It is shown there that only relatively few trajectories are needed for the space-charge forces, and, even more important, there is no need for iteration when the beams are relatively long and narrow. The opposite case, short squat beams with strong longitudinal space-charge forces, occurs in diodes and close to electron-emitting sources. The diode equations are treated in Section 10.4 for the classical planar case, and in Section 10.5 for radial variation of the anode voltage and current.

Chapter 11 differs in that we report on application of the principles that we have explained in previous chapters to electron guns used in television tubes and other CRTs. Of course designs differ from application to application, but certain features are common. The beam-forming region (BFR), the main lens region (MLR), and the need for a drift region (DR) are elaborated upon in Sections 11.2–11.4. Application of the Helmholtz-Lagrange law (and its phase-space generalization) allows us to estimate beam spread due to thermal effects. As discussed, effects of spherical aberration and space-charge beam spreading allow us to estimate the widening of a beam spot due to these effects over and above the thermal widening due to the spread in direction and energy of electrons at the cathode. The problem of spot focussing onto a screen is handled in Section 10.5. We show in Section 10.5.2 that a phenomenological theory of best focus can be produced in many ways (we use only one) to combine the above-mentioned beam spreads into an optimal focussing strategy. The optimization may require redesign of the main lens to lower the aberration coefficient (Section 11.5.3), or one may have to modify the beam-forming region (Section 11.5.4). However, when all is said and done in design of the BFR and ML, one still needs to know the relationship between expected current and voltages to be impressed upon the relevant electron gun grids (the G1 is usually "driven" with respect to the cathode, or vice versa). Cathode emission is introduced in Section 11.6, and the drive characteristics are handled in Section 11.6.1. We show in Section 11.6.2 how a Laplace field calculation allows us to estimate the cathode emission current at given drive voltage even in the presence of strong space-charge forces.

We generally follow the rationalized MKS or SI system of units, but in certain places an older engineering custom of measuring small distances in terms of mils (1 mil = 0.001 inch) is used for convenience. Conversion to millimeters (1 mm ≈ 39.37 mils) is not difficult. In our notation, we use vector $\vec{r}$ to indicate a general space coordinate in three dimensions. We use the vector $\vec{\rho}$ to indicate the transverse Cartesian variables x, y, or the vector $(x, y, 0)$, when the z coordinate is regarded as the principal direction of interest (i.e., the axis of a beam). Derivatives with respect to z are indicated by a prime ($'$); those with respect to a generalized direction coordinate $s = vt$ with an inverse prime ($`$).

The text is written at a level compatible with a four-year degree in physics and/or electrical engineering, possibly with some rudimentary knowledge of how to solve simple total differential equations. The only partial differential equation discussed in the text is the Poisson equation and its cousin, the Laplace equation. The reader need know only that solution to these equations requires values of the unknown field or of its normal derivative on a closed surface for uniqueness. The text attempts to fill in as many gaps as would be compatible with a smoothly flowing exposition without leaving much to the reader to find elsewhere. In this respect it may differ appreciably from existing texts, and the author hopes that this feature will make it more suitable as a textbook for the uninitiated than other texts.

REFERENCES

de Wolf, D.A. (1982) *RCA Rev.*, **43**, 339.

Grivet, P. (1972) *Electron Optics*, 2nd edn., Pergamon, Oxford.

Hall, C.E. (1953) *Introduction to Electron Microscopy*, 2nd edn., McGraw-Hill, New York.

Moss, H. (1968) Narrow angle electron guns and cathode ray tubes, in *Advances in Electronics and Electron Physics* (Supplement 3), Academic Press, New York.

Szilagyi, M. (1988) *Electron and Ion Optics*, Plenum Press, New York.

Wollnik, H. (1987) *Optics of Charged Particles*, Academic Press, Orlando, FL.

CHAPTER 2

THE RELATIONSHIP BETWEEN ELECTRON OPTICS AND LIGHT OPTICS

In the classical, i.e., nonquantum-mechanical, description, electron optics deals with the propagation of point-like charged particles forming a beam in the presence of electromagnetic fields which accelerate, deflect, and/or focus the beam. The wave-like properties of electrons, first described by de Broglie in connection with typical quantum-mechanical phenomena such as the diffraction of electrons through slits, will not be needed in the great majority of applications. Electrons and other charged particles will be treated as point particles with mass, charge, precise location, and precise velocity. Their behavior in an electromagnetic field is sufficiently accurately described in typical electron optics situations by Newton's laws for the acceleration of mass in a force field, where the forces are given (and possibly induced by) electromagnetic fields.

It was realized very early in the development of electron optics (Knoll and Ruska 1932) that the physics of electron beams in electromagnetic fields was very similar to that of light beams in a system of lenses. This section is devoted largely to a discussion of that analogy: the laws that govern the propagation of charged-particle beams in electromagnetic fields are identical to those that govern the propagation of light rays in a medium with a variable refractive index. It will become clear that the particle velocity v plays the role of the refractive index n, and that the trajectories of the particles are identical to the rays of geometrical optics. As a consequence, it is not necessary to establish the manifold lensing results of geometrical optics again in electron optics: there is a large body of results from light optics that carry over directly into electron-optical analogies. However, we will derive many of these results separately for didactical purposes, and also because electron optic lenses are quantitatively so different from the glass lenses of light optics. The latter, for example, usually have fixed and uniform refractive indices, whereas the former are characterized by continuously varying refractive indices in the lensing region.

2.1 GEOMETRICAL OPTICS

Although the reader can find a more extensive treatment in many available optics textbooks, we believe it will be useful to summarize briefly here the essentials of geometrical (or "ray") optics in preparation for demonstrating the analogy between light optics and electron optics.

Many electromagnetic signals are characterized by having a very narrow bandwidth $\delta\omega$ of frequencies around a high carrier angular frequency ω. Such "quasi-monochromatic" signals can be characterized to good approximation by assuming that the time dependence of the complex electric field is given in phasor form by the factor $\exp(-i\omega t)$ multiplying the spatially dependent part of the field. Let such a signal propagate in a linear dispersive medium such as the atmosphere, which can be characterized by the vacuum permeability μ_0 and by a relative permittivity $\epsilon(\vec{r}, \omega)$ that can be a function of frequency. Each frequency component $\vec{E}(\vec{r}, \omega)$ of the signal's electric field vector obeys a vector wave equation,

$$\vec{\nabla} \times \vec{\nabla} \times \vec{E}(\vec{r}, \omega) - k_0^2 n^2(\vec{r}, \omega)\vec{E}(\vec{r}, \omega) = 0 , \tag{2.1}$$

in source-free regions. We define $k_0 \equiv \omega/c$ as the free-space wavenumber, and $n(\vec{r}, \omega)$ as the refractive index, which is the square root of the relative dielectric permittivity $\epsilon(\vec{r}, \omega)$. The refractive index describes the interaction of the wave at given frequency with the medium, and it is all that one needs to know about the medium in geometrical optics. The fact that $\vec{\nabla} \times \vec{\nabla} \times \vec{E} = \vec{\nabla}\vec{\nabla} \cdot \vec{E} - \Delta\vec{E}$, and that $\vec{\nabla} \cdot \vec{E} \neq 0$ in general, signifies that the first term in Eq. (2.1) contains a *polarization-mixing* term, i.e., that E_x, E_y, and E_z, will all occur in each of the three components of the vector equation (2.1). In many optical applications we may approximate $\vec{\nabla} \cdot \vec{E} = 0$, and Eq. (2.1) simplifies into three *scalar* wave equations (one for each Cartesian component),

$$\Delta\vec{E}(\vec{r}, \omega) + k_0^2 n^2(\vec{r}, \omega)\vec{E}(\vec{r}, \omega) = 0 . \tag{2.2}$$

There are obviously a number of conditions which restrict the validity of this equation. It especially requires that the refractive index $n(\vec{r}, \omega)$ varies very little in a wavelength. Even then, care must be exercised because polarization effects can build up cumulatively over large distances. The next point is that each monochromatic component of the quasi-monochromatic signal can be written as a product of a real amplitude $A(\vec{r}, \omega)$ and an exponential factor with a real phase

$$E(\vec{r}, \omega) = A(\vec{r}, \omega) \exp[ik_0 S(\vec{r}, \omega)] . \tag{2.3}$$

The *eikonal* function $S(\vec{r}, \omega)$ has the dimension of length (it reduces to pathlength when $n = 1$) and it plays a prominent part in the following development.

Substitution of Eq. (2.3) into (2.2) yields two equations in the real quantities $A = A(\vec{r}, \omega)$ and $S = S(\vec{r}, \omega)$ (we simplify the notation by leaving off the dependence upon the arguments):

$$n^2 - (\vec{\nabla} S)^2 + k^{-2} \frac{\Delta A}{A} = 0 , \qquad \vec{\nabla} \cdot (A^2 \vec{\nabla} S) = 0 . \tag{2.4}$$

A second condition for validity of the geometrical optics approximation is that the third term in the first of Eqs. (2.4) can be neglected, i.e., that $\Delta A / A \ll \lambda^{-2}$, i.e., that the radius of curvature associated with change in amplitude is much greater than the wavelength. This results in the *eikonal equation*

$$(\vec{\nabla} S)^2 = n^2 . \tag{2.5}$$

The eikonal equation leads to an equation for rays. Planes of constant S are also planes of constant phase, and therefore the gradient of S is in the direction of maximal phase change. A line which has $\vec{\nabla} S$ tangential at every point is known as a *ray*, and rays are the quantities of primary interest in geometrical optics. Once rays have been traced, one can form the *wave fronts* (or planes of constant phase) to reconstruct the electric field E. The ray equation can then be obtained from the eikonal equation by defining a path parameter s and noting that $\vec{\nabla} S = n\hat{s} = n\, d\vec{r}/ds$ by definition. Straight-forward manipulation of this latter equation can be shown (e.g. see Born and Wolf 1965) to lead to

$$\frac{d}{ds}\left(n \frac{d\vec{r}}{ds}\right) = \vec{\nabla} n , \tag{2.6}$$

and one is reminded that the condition leading to Eq. (2.5) holds as well as the condition for ignoring polarization, namely $\lambda|\vec{\nabla}n/n| \ll 1$. Almost all of the geometrical ray-tracing formulas (including Snell's law) are obtained from the ray equation (2.6); the reader will be able to find these in any modern text on geometrical optics.

2.2 FERMAT'S PRINCIPLE AND EULER EQUATIONS

Although we have already sketched a derivation of the ray equation (2.6), we will now repeat it from a different point of view, namely from a variational principle known as *Fermat's Principle*. The sole purpose of doing that in a textbook on electron optics is to make clear the connection between the optics of electrons and those of electromagnetic waves (or light).

Fermat's principle states that light rays will follow the shortest possible path in a medium characterized by a refractive index $n(\vec{r})$. The mathematical statement of this principle is

$$\delta \int_{s_0}^{s_f} ds\, n(s) = 0 , \tag{2.7}$$

where s is a path parameter that we equate with the ray parameter introduced above, and in which expression the endpoints s_0 and s_f are kept fixed under the variations of path implied by the δ preceding the integral sign. Now let the z axis of a Cartesian coordinate system play a special role: $\vec{\rho}(z)$ is then the vector in a $z =$ constant plane that describes any off-axis point. The choice of a z axis is somewhat arbitrary, but of course one will wish to choose it so that it coincides as closely as possible with an idealized optical axis along which a "straightened" beam of rays would propagate. Thus, path point s is represented by the Cartesian coordinates $\vec{\rho}(z)$, z. Simple application of Pythagoras' theorem to the differential elements ds, dz, and $d\vec{\rho}$ yields,

$$ds = \sqrt{d\vec{\rho}^{\,2} + dz^2} = \sqrt{1 + \vec{\rho}\,'^2(z)}\; dz . \tag{2.8}$$

Here $\vec{\rho}\,'(z) \equiv d\vec{\rho}(z)/dz$, and Eq. (2.7) then takes on the form,

$$\delta \int_{z_0}^{z_f} dz\, \mathscr{L}(\vec{\rho}, \vec{\rho}\,', z) = 0 \quad \text{with} \quad \mathscr{L}(\vec{\rho}, \vec{\rho}\,', z) \equiv n(\vec{\rho}, z)\sqrt{1 + \vec{\rho}\,'^2(z)} \tag{2.9}$$

The newly introduced function $\mathscr{L}(\vec{\rho}, \vec{\rho}\,', z)$ is known in classical mechanics as the *Lagrangian*, and we shall see that this is precisely what will make the connection with electron optics. Let us vary the Lagrangian with respect to the variables $\vec{\rho}$ and $\vec{\rho}\,'$ under the restraint of keeping the endpoints fixed:

$$\delta\mathscr{L}(\vec{\rho}, \vec{\rho}\,', z) = \frac{\partial\mathscr{L}}{\partial\vec{\rho}} \cdot \delta\vec{\rho} + \frac{\partial\mathscr{L}}{\partial\vec{\rho}\,'} \cdot \delta\vec{\rho}\,' . \tag{2.10}$$

Because variation and integration can be interchanged, the insertion of this variation into Eq. (2.9) yields

$$\int_{z_0}^{z_f} dz \left(\frac{\partial\mathscr{L}}{\partial\vec{\rho}} \cdot \delta\vec{\rho} + \frac{\partial\mathscr{L}}{\partial\vec{\rho}\,'} \cdot \delta\vec{\rho}\,' \right) = 0 . \tag{2.11}$$

Replace $\delta\vec{\rho}\,'$ by $d(\delta\vec{\rho})/dz$, and integrate the second term by parts to obtain for any z_0 and z_f,

$$\int_{z_0}^{z_f} dz \left[\frac{\partial\mathscr{L}}{\partial\vec{\rho}} - \frac{d}{dz}\left(\frac{\partial\mathscr{L}}{\partial\vec{\rho}\,'}\right)\right] \cdot \delta\vec{\rho} = 0 \rightarrow \frac{\partial\mathscr{L}}{\partial\vec{\rho}} - \frac{d}{dz}\left(\frac{\partial\mathscr{L}}{\partial\vec{\rho}\,'}\right) = 0 . \tag{2.12}$$

The last equation in (2.12) follows because the variation is arbitrary with respect to the vector $\delta\vec{\rho}$, and it is actually a pair of *Euler equations*; one for each transverse coordinate. The Euler equations lead directly to the equations for geometrical optics rays. To see how, we invoke the definition of $\mathscr{L}$ in Eq. (2.9) to observe that

$$\frac{\partial\mathscr{L}}{\partial\vec{\rho}} = \sqrt{1 + \vec{\rho}\,'^2}\, \vec{\nabla}_T n , \qquad \vec{\nabla}_T \equiv \hat{x}\frac{\partial}{\partial x} + \hat{y}\frac{\partial}{\partial y} ,$$

$$\frac{\partial\mathscr{L}}{\partial\vec{\rho}\,'} = \frac{\vec{\rho}\,'}{\sqrt{(1 + \vec{\rho}\,'^2)}}\, n = \frac{n\, d\vec{\rho}}{\sqrt{(1 + \vec{\rho}\,'^2)}\, dz} = n\frac{d\vec{r}}{ds} .$$

Insert these expressions for derivatives of $\mathscr{L}$ into Eq. (2.12) to obtain

$$\sqrt{1 + \vec{\rho}\,'^2}\, \vec{\nabla}_T n - \frac{d}{dz}\left(n\frac{d\vec{\rho}}{ds}\right) = 0 .$$

Divide by $\sqrt{1 + \vec{\rho}\,'^2}$ to obtain

$$\vec{\nabla}_T n - \frac{d}{ds}\left(n\frac{d\vec{\rho}}{ds}\right) = 0 . \tag{2.13}$$

This equation represents two of the three equations implied in vector form by Eq. (2.6), and we obtain it directly by taking the dot product of that equation with the unit vector $\hat{\rho} = (\hat{x} + \hat{y})/\sqrt{2}$. The third equation, obtained from the dot product of Eq. (2.6) with $\hat{z}$, becomes a trivial identity if we identify the local $\hat{z}$ direction with that of the path parameter unit vector $\hat{s}$ at the location in question. As the choice of $\hat{z}$ is arbitrary, the transverse-coordinate equation (2.13) is essentially identical to Eq. (2.6), and we have merely derived a special form of the ray equation from Fermat's principle. If we choose $\hat{z}$ different from $\hat{s}$, then it can be shown with a bit more work (see, e.g., Marcuse 1972), that

$$\frac{\partial n}{\partial z} - \frac{d}{ds}\left(n \frac{dz}{ds}\right) = 0 . \tag{2.14}$$

also follows from Eqs. (2.13), which therefore suffice to describe the propagation of rays. That two transverse equations suffice is also clarified above by the possibility of choosing $\hat{z} = \hat{s}$ locally.

2.3 PRINCIPLE OF LEAST ACTION

Now we switch our attention from physical and geometrical optics of waves to the physics and optics of point particles which, in the macroscopic world, are governed by Newton's laws (at non-relativistic velocities) and their consequences. Point particles suggest microscopic physics which in turn requires a quantum-mechanical description, and the phenomena of moving particles are described properly only by the twentieth century theories of relativity, but almost all electron optics at low particle kinetic energies can be given to adequate accuracy by Newtonian physics coupled with Maxwell's equations. The starting point for us is the Newtonian law that the force on a particle is proportional to its acceleration, and the constant of proportionality is the mass of the particle. In electron optics, the particle has a charge and is therefore susceptible to electromagnetic forces. We will neglect the gravitational forces of charged particles upon each other (and those due to the Earth, etc.) as this force is orders of magnitude less than the electromagnetic forces produced by moving charges or by laboratory situations. The force upon a point particle with mass m and electric charge q ($q = -e$ for electrons, with $e \approx 1.602 \times 10^{-19}$ Coulombs) is $\vec{F} = q\vec{E}$, if $\vec{E}$ is the electric field-strength vector at the location of the particle. The Newtonian law then states,

$$m\ddot{\vec{r}} = -q\vec{\nabla}\Phi \quad \text{with} \quad \ddot{\vec{r}} = \frac{d^2\vec{r}}{dt^2} , \tag{2.15}$$

because we may write $\vec{E} = -\vec{\nabla}\Phi$ in electrostatics: the field-strength vector can always be expressed as the gradient of a scalar *potential* field Φ. There will be occasions on which we will have to modify Eq. (2.15) to incorporate relativistic-velocity effects, but that will come at a later stage of this development. Classical physics (i.e., physics that excludes quantum phenomena at the microscopic level) requires merely that an initial position $\vec{r}_0$ and an initial velocity $\dot{\vec{r}}_0 = (d\vec{r}/dt)_{\vec{r}=\vec{r}_0, t=t_0}$ of each particle be specified. Knowledge of the electrostatic potential $\Phi(\vec{r})$ is then sufficient for obtaining the solution of Eq. (2.15) for the particle trajectories in an electrostatic situation in which Φ is not dependent upon time for the duration in which the charged particles produce their trajectories. If $\Phi(\vec{r}, t)$ is time-dependent, then the fields are not stationary and the laws of electrostatics, especially $\vec{E} = -\vec{\nabla}\Phi$, no longer hold. At some point, we shall expand Eq. (2.15) to include the effect of magnetic forces, but at this stage we shall assume that electrostatics holds at least adiabatically, i.e., so that time changes in $\Phi(\vec{r}, t)$ occur so slowly that we may ignore electrodynamic terms in Maxwell's equations (the $\partial\vec{B}/\partial t$ term in the curl-E equation and the $\partial\vec{D}/\partial t$ term in the curl-H equation).

For wave optics we outlined a direct physical-optics derivation of the ray equations (2.6) and (2.13). We then rederived the very same ray equations from a variational principle, Eq. (2.7). Following a similar procedure, we will start from a variational principle of particle physics which is equivalent to Eq. (2.15), and use that to derive ray equations for electron optics. To do so, we define the *Lagrangian* function $\mathcal{L}(\vec{r}, \dot{\vec{r}}, t) \equiv T(\vec{r}, \dot{\vec{r}}, t) - U(\vec{r}, \dot{\vec{r}}, t)$; is is the difference between kinetic energy T and potential energy U. For a single charged particle we have

$$T = \tfrac{1}{2} m\dot{\vec{r}}^2 , \qquad U = q\Phi . \tag{2.16}$$

The reader is reminded that $\vec{F} = -\vec{\nabla}U$, which yields the electrostatic force $\vec{F} = q\vec{E}$ and field $\vec{E} = -\vec{\nabla}\Phi$. Thus the Lagrangian for a charged particle in an electrostatic field is

$$\mathcal{L}(\vec{r}, \dot{\vec{r}}, t) = \tfrac{1}{2} m\dot{\vec{r}}^2 - q\Phi(\vec{r}) , \tag{2.17}$$

and the equivalent variational principle is

$$\delta \int_{t_0}^{t_f} dt\, \mathcal{L}(\vec{r}, \dot{\vec{r}}, t) = 0 . \tag{2.18}$$

The Euler equations for $\mathcal{L}$ follow from this principle by exactly the same steps taken in connection with Eqs. (2.9)–(2.12), provided we replace

$(\vec{\rho}, \dot{\vec{\rho}}, z)$ by $(\vec{r}, \dot{\vec{r}}, t)$, and with similar replacements for derivatives. We then obtain

$$\frac{\partial \mathcal{L}}{\partial \vec{r}} - \frac{d}{dt}\left(\frac{\partial \mathcal{L}}{\partial \dot{\vec{r}}}\right) = 0 . \tag{2.19}$$

We have three equations here, of course, e.g. one for each of the three Cartesian coordinates. When the Lagrangian of Eq. (2.17), which has no explicit dependence upon time t, is applied to Eq. (2.19), we immediately retrieve the Newtonian equation of motion (2.15), which establishes the variational principle (2.18) as a valid point of departure. On the other hand, Eq. (2.18) can be made to resemble Eq. (2.7) even more closely if we succeed in somehow transforming from time coordinate t to path coordinate s. For conservative forces (i.e., those that can be considered as the gradient of a potential field), the *Hamiltonian* function $\mathcal{H}(\vec{r}, \dot{\vec{r}}, t) \equiv T(\vec{r}, \dot{\vec{r}}, t) + U(\vec{r}, \dot{\vec{r}}, t)$ is conserved along all particle path (or trajectory) locations s. Hence, if the initial points t_0, s_0 and the final points t_f, s_f are kept fixed (as is necessary in order to obtain Euler's equations) then $\mathcal{H}$ is constant in the variations implied by Eq. (2.18). We may thus set $\mathcal{L} = T - U = 2T - \mathcal{H}$ in Eq. (2.18), and then use the fact that $\delta\mathcal{H} = 0$ to obtain

$$\delta \int_{t_0}^{t_f} dt\, \mathcal{L}(\vec{r}, \dot{\vec{r}}, t) = 2\delta \int_{t_0}^{t_f} dt\, T(\vec{r}, \dot{\vec{r}}, t) = \delta \int_{t_0}^{t_f} dt\, m\dot{\vec{r}}^2 = \delta \int_{s_0}^{s_f} ds\, mv(s) , \tag{2.20}$$

because $\dot{\vec{r}}\, dt = d\vec{s}$, and vector $d\vec{s}$ is parallel to the velocity vector $\vec{v}(s) = \dot{\vec{r}}$ (so that the dot product becomes just a scalar product of two path quantities). The consequence of this is that the variational principle (2.18) transforms into the *principle of least action*:

$$\delta \int_{s_0}^{s_f} ds\, mv(s) = 0 . \tag{2.21}$$

We observe that the boundary values t_0 and t_f transform into s_0 and s_f as time t is transformed into path parameter s. The obvious similarity between Eqs. (2.21) and (2.7) suggests that the *momentum* $mv(s)$ of the particle plays the same role as the *refractive index* $n(s)$ of the medium in wave optics. This analogy is further strengthened by noting that Eq. (2.15), which follows from the path variation (2.21) with fixed endpoints, can be made to resemble Eq. (2.6). We simply note that

$$\frac{d}{dt} = \frac{ds}{dt}\frac{d}{ds} = v(s)\,\frac{d}{ds}\,, \tag{2.22}$$

and therefore that

$$m\ddot{\vec{r}} = v(s)\,\frac{d}{ds}\left[mv(s)\,\frac{d\vec{r}}{ds}\right] = \vec{F}\,. \tag{2.23}$$

For *conservative* systems we have noted that $\mathcal{H}(s)$ is constant along a trajectory, and thus that $\vec{F} = -\vec{\nabla}U = \vec{\nabla}T$ because $\mathcal{H} = T + U$. Using $T = \frac{1}{2}mv^2(s)$ we then obtain

$$\vec{F} = v(s)\vec{\nabla}[mv(s)]\,. \tag{2.24}$$

Combination of Eqs. (2.23) and (2.24) finally leads to

$$\frac{d}{ds}\left[mv(s)\,\frac{d\vec{r}}{ds}\right] = \vec{\nabla}[mv(s)]\,. \tag{2.25}$$

This formula is identical to Eq. (2.6): in electron optics the momentum $mv(s)$ takes the place of the refractive index $n(s)$ of wave optics. The fact that $mv(s)$ has a dimension is of little relevance because nothing changes in Eq. (2.25) if both sides are divided by mc, where $c \approx 2.998 \times 10^8$ m/s is the velocity of light in vacuo. That simply replaces $mv(s)$ by the dimensionless ratio $v(s)/c$. The essential fact is that the velocity $v(s)$ of the point particle is the analog of the refractive index due to wave interaction with a medium.

2.4 WAVE VERSUS PARTICLE OPTICS

We make use of the conservation of the Hamiltonian $\mathcal{H}(s)$ along a particle trajectory to replace velocity $v(s)$ because it is not a directly measurable quantity in most electron optics situations. Hence, we use $\mathcal{H}(s) = \mathcal{H}(s_0)$, with s_0 as the initial point of the trajectory and $s > s_0$ as a path point. Because $q = -e$ for electrons, this equality yields

$$\begin{aligned} \tfrac{1}{2}mv^2(s) - e\Phi(s) &= \tfrac{1}{2}mv^2(s_0) - e\Phi(s_0) \\ &\equiv \tfrac{1}{2}mv_0^{\,2} - e\Phi_0\,. \end{aligned} \tag{2.26}$$

We will use the abbreviated notation $v \equiv v(s)$ and $\Phi \equiv \Phi(s)$, so that Eq. (2.26) yields a value for the velocity,

$$v = \sqrt{\frac{2e}{m}}\left[\Phi - \Phi_0 + \frac{mv_0^2}{2e}\right]^{1/2} . \qquad (2.27)$$

Note that v is a function of a potential *difference*, $\Phi - \Phi_0$. Electrostatic potentials are always measured with respect to a reference value (often taken to be zero at infinity for useful physical reasons). Here, it is more useful to our ensuing development to choose the reference value Φ_0 such that $e\Phi_0 = \frac{1}{2}mv_0^2$. That implies that $\mathscr{H}(s_0) = 0$, and from Eqs. (2.26) and (2.27) we then have

$$v = \sqrt{\frac{2e\Phi_a}{m}} \quad \text{or} \quad mv = \sqrt{2em\Phi_a}\,, \qquad (2.28)$$

and therefore we see that $v(s)$ is proportional to the square root of the *accelerating potential* $\Phi_a(s)$, i.e., to the value of potential $\Phi(s)$ with respect to a reference value such that $\Phi_0 = mv_0^2/2e$ when that potential is expressed with respect to the same reference value.

In summary, electron optics can be obtained from geometrical optics of waves upon making the substitution,

$$n(s) \rightarrow \kappa\sqrt{\Phi_a(s)}\,, \qquad (2.29)$$

where $\Phi_a(s)$ is the accelerating potential and κ is a constant, possibly with a dimension. We shall reaffirm that the choice of constant κ is unimportant in almost all cases of interest. The analogy implied by Eq. (2.29) enables electron optics designers to calculate many effects of electron beams in electromagnetic fields simply by utilizing results well known in geometrical optics of waves.

It is possible to take the analogy too rigidly because there are also important differences (El-Kareh and El-Kareh 1970). The most important limitations of the analogy are:

(i) Refractive indices in wave optics vary little from one medium to another, and the changes are often discontinuous. In electron optics, the changes are always continuous and they are often very large.

(ii) Light photons exert no forces upon each other (classically), but electrons interact with forces proportional to the inverse square of their separation distance. The effect of these *Coulomb* or *space-charge* forces is very important in many particle-beam situations.

Both wave and particle optics, as defined by the equations involving an effective refractive index $n(s) \leftrightarrow mv(s)$, are subject to certain restrictions in their regions of validity. In both cases we must have

$$\lambda \left| \frac{1}{n} \vec{\nabla} n \right| \ll 1 . \tag{2.30}$$

The meaning of this in wave optics has already been discussed. In particle optics, the corresponding wavelength quantity is the de Broglie wavelength $\lambda = h/mv(s)$, where $h \approx 6.624 \times 10^{-34}$ J/s is Planck's constant. The de Broglie wavelength for electrons is (λ in Å) ≈ 12.26 ($\Phi_a^{-1/2}$ in volts). Hence Eq. (2.30) states that the change in $\Phi_a(s)$ must not be apprecible over distances $\delta s \sim \lambda$. This condition is easily met in almost all particle optics situations because λ is so small.

REFERENCES

Born, M. and E. Wolf (1965) *Principles of Optics*, 3rd edn., Pergamon, Oxford.

El-Kareh, A. B. and J. C. J. El-Kareh (1970) *Electron Beams, Lenses, and Optics*, Academic Press, Orlando, FL.

Knoll, M. and E. Ruska (1932) *Ann. Physik* **12**, 607, 641.

Marcuse, D. (1972) *Light Trasmission Optics*, Van Nostrand Reinhold, New York.

CHAPTER 3

EQUATIONS OF MOTION

In this chapter, we shall take the point of view that the electromagnetic fields $\vec{E}$ and $\vec{B}$ are known everywhere. The purpose of this section is then to determine trajectory equations, for electrons with charge $q = -e \approx -1.602 \times 10^{-19}$ Coulombs, that are determined in turn by a given electromagnetic field. In Chapter 4 we describe the determination of the fields $\vec{E}(\vec{r})$ and $\vec{H}(\vec{r})$ from given charge and current sources, or, alternatively, from given boundary conditions such as conducting plates maintained at fixed potentials. These two aspects of electron optics, the motion of charges and the electromagnetic fields, are interconnected in a very real sense because the presence of a beam of moving charges can induce a charge distribution on nearby conductors which in turn influences the electromagnetic fields that "steer" the beam. Even more pertinent; the charges produce Coulomb forces upon each other which perturb the motion of those charges due to electromagnetic fields produced by exterior sources. In spite of these complications, it is sufficiently often true that these aspects can be considered more or less independently, at least to good approximation. Our treatment will also be restricted to beams of electrons; beams of other charged particles can be treated similarly upon replacing $-e$ by charge q of the particle species.

3.1 NONRELATIVISTIC TRAJECTORY EQUATIONS

In the classical physics description, a point particle with mass m and charge q undergoes a force in an electromagnetic field that is given by the expression

$$\vec{F}_L = q(\vec{E} + \dot{\vec{r}} \times \vec{B}) , \tag{3.1}$$

where $\dot{\vec{r}}$ is the velocity vector of the particle, so that the component of this *Lorentz force* induced by the magnetic flux density is perpendicular at all times to the trajectory (for which $d\vec{s} = \dot{\vec{r}}\, dt$). Newton's law of acceleration states that $\vec{F} = m\ddot{\vec{r}}$. Although the quantum description of matter renders that imprecise at the microscopic level, we do assume here that the particle's location $\vec{r}$ and velocity $\dot{\vec{r}}$ can be given with arbitrary precision at all times, provided initial values $\vec{r}_0$ of location and $\dot{\vec{r}}_0$ of velocity are specified. The trajectory of an electron can be found from Eq. (3.1) in principle if fields $\vec{E}(\vec{r}, t)$ and $B(\vec{r}, t)$ are known at all $\vec{r}$ and t, but the time dependence of these fields can make the procedure of determining the trajectories quite complicated.

We shall restrict ourselves in this section to deriving trajectory equations for a beam of electrons (with charge $q = -e$) under the influence of *static* electromagnetic fields $\vec{E}(\vec{r})$ and $\vec{B}(\vec{r})$. Beams of electrons then have a time-invariant shape, i.e., the local charge density is constant although electrons move in and out of any volume element. The derivative d/dt may be written as $(ds/dt)d/ds = v(s)d/ds$. An alternative way of expressing the time invariance is to note that $dt = ds/v(s)$, and hence that all variations with time translate into motion along fixed paths. No matter what time interval dt is under consideration, $dt = ds/v(s)$ at any path point. It is advantageous to derive trajectory equations with a spatial coordinate rather than time t as independent variable.

Either the path coordinate s of a trajectory or a Cartesian coordinate z (chosen in some preferred direction, e.g., the direction of the "centroid" of the beam) can be chosen as the independent variable. We will have occasion to use either in the ensuing development. We shall assume that the electrostatic forces are all derivable from an electrostatic potential, i.e., that $\vec{E}(\vec{r}) = -\vec{\nabla}\Phi(\vec{r})$. It is easily demonstrated that the Hamiltonian $\mathcal{H}(s)$ is conserved along a trajectory, i.e., that $d\mathcal{H}(s)/ds = 0$. Simply write down the Newtonian equation for the Lorentz force obtained from Eq. (3.1),

$$m\ddot{\vec{r}} = -e(\vec{E} + \dot{\vec{r}} \times \vec{B}) , \tag{3.2}$$

and dot multiply both sides of this equation by the velocity vector $\dot{\vec{r}}$

$$m\dot{\vec{r}} \cdot \ddot{\vec{r}} = -e\dot{\vec{r}} \cdot \vec{E} . \tag{3.3}$$

The left-hand side is rewritten with the aid of

$$\dot{\vec{r}} \cdot \ddot{\vec{r}} = \frac{1}{2} \frac{d}{dt} (\dot{\vec{r}}^2) , \tag{3.4}$$

and for the right-hand side of Eq. (3.3) we note that

$$\frac{d}{dt} \Phi(\vec{r}, t) = \frac{\partial \Phi}{\partial t} + \dot{\vec{r}} \cdot \vec{\nabla}\Phi = \frac{\partial \Phi}{\partial t} - \dot{\vec{r}} \cdot \vec{E} . \tag{3.5}$$

Conservative forces arise from potentials with $\partial\Phi/\partial t = 0$ so that $\dot{\vec{r}} \cdot \vec{E} = -d\Phi/dt$. When this identity is combined with Eqs. (3.4) into (3.3), the result is

$$\frac{d}{dt} [\tfrac{1}{2} m\dot{\vec{r}}^2 - e\Phi(\vec{r})] = 0 , \tag{3.6}$$

and we need only replace the derivative d/dt by $v(s)d/ds$, as set forth above, to rewrite this conservation equation as

$$\frac{d}{ds} [\tfrac{1}{2} m\dot{\vec{r}}^2 - e\Phi(\vec{r})] = 0 \quad \text{or} \quad \frac{d\mathcal{H}(s)}{ds} = 0 . \tag{3.7}$$

We recognize the expression in square brackets in the first of Eqs. (3.7) as the Hamiltonian function $\mathcal{H}(s)$. Both $\vec{r}(s)$ and $\dot{\vec{r}}(s)$ are considered as functions of the path parameter in the second expression. Equations (2.26)–(2.28), combined with (3.7), express the fact that total energy is conserved, i.e., that

$$\tfrac{1}{2} m\dot{\vec{r}}^2 = e\Phi_a(\vec{r}) , \tag{3.8}$$

and this is obtained explicitly by setting $\mathcal{H}(s) = \mathcal{H}(s_0)$ and then defining the *accelerating potential*,

$$\Phi_a(\vec{r}) \equiv \Phi(\vec{r}) - \Phi(\vec{r}_0) + \frac{m\dot{\vec{r}}^2}{2e} . \tag{3.9}$$

Let us now introduce a Cartesian coordinate system with transverse coordinate vector $\vec{\rho} = (x, y, 0)$ so that z is the preferred direction. We can then rewrite Eq. (3.8) as

$$\tfrac{1}{2} m(\dot{\vec{\rho}}^2 + \dot{z}^2) = e\Phi_a(\vec{\rho}, z) . \tag{3.10}$$

In this coordinate system, we describe the trajectory by the evolution of the transverse coordinate $\vec{\rho}(z)$, and no further independent coordinate is needed if the fields are static. We shall see shortly that the above description in terms of a Cartesian z coordinate is not always possible, but if it is, then the equation we seek will be in terms of derivatives of $\vec{\rho}(z)$ with respect to z. We shall use a prime notation for derivatives with respect to z, i.e., $\vec{\rho}'(z) \equiv d\vec{\rho}/dz$ and $\vec{\rho}''(z) \equiv d^2\vec{r}/dz^2$. Some algebraic manipulation is now needed to obtain the desired trajectory equations. First consider $\dot{\vec{\rho}}$:

$$\dot{\vec{\rho}} \equiv \frac{d\vec{\rho}}{dt} = \frac{dz}{dt}\frac{d\vec{\rho}}{dz} \equiv \dot{z}\vec{\rho}' \ . \tag{3.11}$$

Use this in Eq. (3.10) to eliminate $\dot{\vec{\rho}}^2$ and so obtain

$$\dot{z}^2 = \frac{2e\Phi_a}{m}\left(\frac{1}{1+\vec{\rho}'^2}\right) . \tag{3.12}$$

Then take the time derivative of Eq. (3.11):

$$\ddot{\vec{\rho}} = \frac{d}{dt}(\dot{z}\vec{\rho}') = \ddot{z}\vec{\rho}' + \dot{z}\frac{d}{dt}(\vec{\rho}') = \ddot{z}\vec{\rho}' + \dot{z}^2\vec{\rho}'' \ . \tag{3.13}$$

Eliminate $\dot{z}^2$ from Eqs. (3.12) and (3.13) to obtain

$$\vec{\rho}'' = \frac{m}{2e\Phi_a}(1+\vec{\rho}'^2)(\ddot{\vec{\rho}} - \vec{\rho}'\ddot{z}) \ . \tag{3.14}$$

Use the three components of the vector equation (3.2) to obtain from Eq. (3.14) the following two equations,

$$x'' = -\frac{1+\vec{\rho}'^2}{2\Phi_a}\{(E_x - x'E_z) + [x'\dot{y}B_x - (\dot{z} + x'\dot{x})B_y + \dot{y}B_z]\} ,$$
$$y'' = -\frac{1+\vec{\rho}'^2}{2\Phi_a}\{(E_y - y'E_z) + [(\dot{z} + y'\dot{y})B_x - y'\dot{x}B_y - \dot{x}B_z]\} . \tag{3.15}$$

It remains only to remove the three time derivatives of Cartesian cordinates in these equations. To do that, it is noted that $\dot{x} = \dot{z}x'$ and $\dot{y} = \dot{z}y'$. If these two are substituted into Eq. (3.15), an expression ensues that contains time derivatives only in the combinations $\dot{z}B_x$, $\dot{z}B_y$, and $\dot{z}B_z$. Let us abbreviate B_i for any of the three Cartesian components of $\vec{B}$, so that

$$\dot{z}B_i = \left(\frac{m\dot{z}^2}{e}\right)\left(\frac{eB_i}{m\dot{z}}\right) . \tag{3.16}$$

The vector $e\vec{B}/m\dot{z} \equiv \vec{K}$ has the dimension of inverse length l^{-1}, and the inverse of its transverse component, K_ρ^{-1}, is known as the *cyclotron radius* (see Section 5.4) because it describes the radius of the circular motion that B_ρ produces upon a charged particle, with initial velocity $v_z = \dot{z}$, in a plane through $\hat{\rho}$ and the z axis (an electron feels the force $-e\vec{v}_z\hat{z} \times \vec{B}$ in this plane). With Eq. (3.12), we then obtain

$$\dot{z}B_i = \frac{2\Phi_a}{(1+\vec{\rho}'^2)} K_i \,. \tag{3.17}$$

Hence if we first insert $\dot{x} = \dot{z}x'$ and $\dot{y} = \dot{z}y'$ into Eq. (3.15), and subsequently replace all $\dot{z}B_i$ factors by means of Eq. (3.17) we obtain

$$x'' = -\frac{1+\vec{\rho}'^2}{2\Phi_a}(E_x - x'E_z) - [x'y'K_x - (1+x'^2)K_y + y'K_z]\,,$$
$$y'' = -\frac{1+\vec{\rho}'^2}{2\Phi_a}(E_y - y'E_z) + [-(1+y'^2)K_x + x'y'K_y + x'K_z]\,, \tag{3.18}$$

with $\vec{\rho}'^2 \equiv x'^2 + y'^2$ and $\vec{K} \equiv e\vec{B}/mv_z$. The two equations can be combined more succinctly into a single vector equation:

$$\vec{\rho}'' = -\frac{1+\vec{\rho}'^2}{2\Phi_a}(\vec{E}_\rho - \vec{\rho}'E_z) + \hat{z} \times [\vec{r}' \times (\vec{r}' \times \vec{K})]\,. \tag{3.19}$$

It is understood that Φ_a, $\vec{E}_\rho$, E_z, and $\vec{B}$ in these equations can all be functions of $\vec{r}$. To solve these trajectory equations for an electron in static electromagnetic fields, we need to know x_0, y_0, and x_0', y_0' at an initial point s_0 for each trajectory. At any location $s > s_0$, we can obtain $\dot{z} = v_z(s)$ by means of Eq. (3.12) from our knowledge of $\vec{\rho}'(z)$ and $\Phi_a(z)$. That enables us to specify $\vec{K}(z)$ and then to use Eqs. (3.18) or (3.19) to obtain $\vec{\rho}(z + dz)$ and $\vec{\rho}'(z + dz)$ by a "marching" method of solving a coupled set of ordinary differential equations (see Section 3.6 for more details).

While this procedure may seem straightforward, it can stumble on a snag sketched in Fig. 3-1. The sketch shows a trajectory with path coordinate $s(z)$ that gradually bends away from the z axis. At point s_t (a *turning point*), we observe that ρ' has become infinite and that the function $\rho(z)$ is not single-valued beyond that path point. In fact, $\rho(z)$ is *multivalued* for $z \leqslant z_t$. The numerical solution of Eq. (3.19), advanced from $s = s_0$, will fail when the turning point s_t is reached! In such situations, i.e., situations in which turning points render a description of $\vec{\rho}$

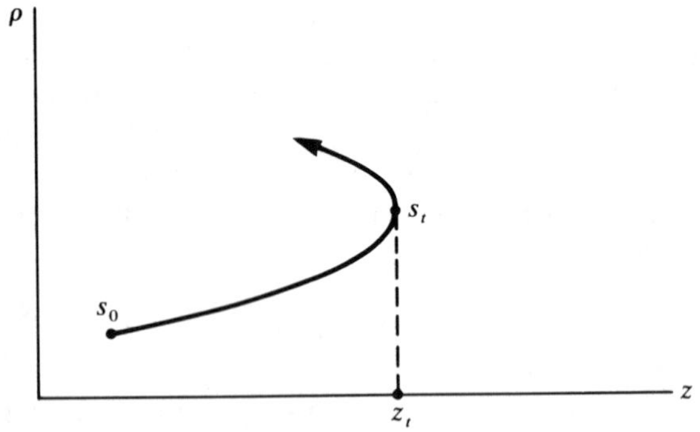

Fig. 3-1. An electron trajectory with a turning point.

in terms of z less useful, it is preferable to take $s = |\dot{\vec{r}}|t = vt$ as the independent coordinate. In the following, we will denote derivatives with respect to path parameter s by prime symbols that slant in the opposite direction, i.e.,

$$\frac{d\vec{\rho}}{dz} \equiv \vec{\rho}\,'(z)\,, \qquad \frac{d\vec{\rho}}{ds} \equiv \vec{\rho}\,`(s)\,, \tag{3.20}$$

We will now develop equations for $\vec{\rho}\,``(s)$ that resemble Eq. (3.19), and it will become clear that the above difficulty due to multivaluedness of $\vec{\rho}(z)$ does not occur in this formulation. First, Eq. (3.8) can be applied to replace velocity $v(s)$ at trajectory point s by the accelerating potential:

$$v(s) = \sqrt{\frac{2e\Phi_a(s)}{m}}\,. \tag{3.21}$$

Using the differential relationship $ds = v(s)\,dt$, it follows that the derivative d/dt can be replaced by $v(s)d/ds$, and therefore that the preceding equations yield,

$$\begin{aligned} \dot{\vec{\rho}} &= v(s)\,\frac{d\vec{\rho}}{ds} \equiv v\vec{\rho}\,`\,, \\ \ddot{\vec{\rho}} &= v(s)\,\frac{d}{ds}\,(v(s)\vec{\rho}\,`) = v^2\vec{\rho}\,`` + vv`\vec{\rho}\,` \\ &= v^2\vec{\rho}\,`` + \tfrac{1}{2}\vec{\rho}\,`\,\frac{d}{ds}\,(v^2)\,. \end{aligned} \tag{3.22}$$

We eliminate v^2 by again referring to Eq. (3.21), to obtain

$$\ddot{\vec{\rho}} = \frac{2e}{m} (\Phi_a \vec{\rho}'' + \tfrac{1}{2}\Phi_a' \vec{\rho}') . \tag{3.23}$$

Combine this equation with the transverse component of the Newtonian force equation (3.2). Upon elimination of $\ddot{\vec{\rho}}$, we obtain

$$\vec{\rho}'' = -\frac{1}{2\Phi_a} [\vec{E}_\rho + v(\vec{r} \times \vec{B})_\rho] - \frac{\Phi_a'}{2\Phi_a} \vec{\rho}' . \tag{3.24a}$$

Precisely the same steps can be followed from Eq. (3.22) to (3.24a) for $\dot{z}$, $\ddot{z}$, and the z component of Eq. (3.2), to obtain

$$z'' = -\frac{1}{2\Phi_a} [\vec{E}_z + v(\vec{r} \times \vec{B})_z] - \frac{\Phi_a'}{2\Phi_a} z' . \tag{3.24b}$$

The last two equations can be combined into

$$\vec{r}'' = -\frac{1}{2\Phi_a} [\vec{E} + v(\vec{r} \times \vec{B})] - \frac{\Phi_a'}{2\Phi_a} \vec{r}' . \tag{3.24c}$$

Finally, we use the fact that

$$\Phi_a' \equiv \frac{d\Phi_a}{ds} = \frac{d\Phi}{ds} = \frac{\partial \Phi}{\partial z} \frac{dz}{ds} = -E_z z' , \tag{3.25}$$

to modify the last term of Eq. (3.24c), and a rearrangement of terms then leads to

$$\vec{r}'' = -\frac{1}{2\Phi_a} [E - \vec{r}' z' E_z] - (\vec{r}' \times \vec{K}) . \tag{3.26}$$

The reader is reminded that $\vec{K} = e\vec{B}/m\dot{z}$. This equation, in contrast to Eq. (3.19), expresses the trajectory in terms of the path coordinate $s = vt$. It appears to be slightly simpler in form than Eq. (3.19), but we have three partial differential equations of second order here, as opposed to two when z is the independent coordinate. From $\dot{\vec{r}} = v\vec{r}'$ we observe that $\vec{r}' = \vec{v}/v = \hat{s}$: the unit vector tangential to the trajectory at point s. Likewise we observe that $z' = v_z/v$ and that $\vec{\rho}' = \vec{v}_\rho/v$, i.e., that the three derivatives of x, y, z with respect to path coordinate s are simply the direction cosines with respect to the chosen Cartesian coordinate system. The vector equation (3.26) can be solved by a "marching" method from s to $s + ds$, provided initial conditions $\vec{r}(s_0)$ and $rv'(s_0)$ are given for each trajectory. This is possible only, of course, if the fields $\vec{E}(s)$ and $\vec{B}(s)$ are specified everywhere. The "marching" method then advances $\vec{r}(s)$ and

$\vec{r}\,'(s)$ to $\vec{r}(s + ds)$ and $\vec{r}\,'(s + ds)$ for values of ds sufficiently small; various numerical integration schemes differ in details as to how this is done. Unfortunately, the fields are not fully known when "space-charge" effects (i.e., Coulomb forces between the charged particles) are strong because the trajectory at s will depend upon Coulomb fields due to charged particles beyond s from this and other trajectories. Such problems will be treated later, and approximations will be brought forward in Chapter 10 that allow us to continue making the assumption that the fields at s are fully known for the purpose of advancing a numerical solution to $s + ds$.

Let us make a comparison between the two sets of equations, (3.19) in terms of the z coordinate, and (3.26) in terms of the s coordinate as independent parameter. The former fail at a turning point because $\vec{\rho}\,'(z) \equiv d\vec{\rho}/dz \rightarrow \infty$ there. On the other hand, we see that $\vec{\rho}\,'(s) \equiv d\vec{\rho}/ds \rightarrow \hat{\rho}$ and $z'(s) \equiv dz/ds \rightarrow 0$ at the turning point; hence Eq. (3.26) poses no singularity problem at turning points. Even so, we prefer to use Eq. (3.19) whenever possible because equipotential surfaces often lie on planes outside of lensing or focussing regions, and even in lensing regions it is advantageous to be able to know the trajectory coordinates on a plane perpendicular to a main axis direction. On the other hand, if we use $s = vt$ as the independent coordinate, then the loci of constant s for trajectories in a beam may not coincide with a useful surface upon solving trajectories with Eq. (3.26). Nevertheless, the problem of turning points, e.g., as would be present with an electron optics mirror, might necessitate the use of that equation.

3.2 NONRELATIVISTIC MOMENTUM FORMULATION

It is useful for many applications, e.g., for periodic electron lenses, to formulate the trajectory equations in terms of momentum rather than velocity or direction cosines. One reason for this is that we can view the motion of a particle as a transformation of position variable $\vec{r}$ and momentum variable $\vec{p}$ into new values along the trajectory, and a convenient matrix algebra can be set up to describe the motion through all kinds of lenses (see Chapter 9). At present, we introduce the new formulation by starting from the ray-trajectory coordinates $\vec{\rho}(z)$ and $\vec{\rho}\,'(z) \equiv d\vec{\rho}/dz$ (the direction tangent vector at z). This is slightly less general than a description in terms of the curvilinear coordinate $s = vt$, but it is more convenient for the applications we have in mind, and the more general description can be handled in a similar fashion.

The transverse particle momentum is $\vec{p}_T = m\dot{\vec{\rho}}(z) = m\dot{z}\vec{\rho}\,'(z)$, and the longitudinal momentum is $p_z = m\dot{z}$. We eliminate $\dot{z}$ with the aid of Eq. (3.12):

$$\vec{p}_T = \sqrt{2em\Phi_a}\,\frac{\vec{\rho}\,'}{\sqrt{(1+\vec{\rho}\,'^2)}}\,, \qquad p_z = \sqrt{2em\Phi_a}\,\frac{1}{\sqrt{(1+\vec{\rho}\,'^2)}}\,. \tag{3.27}$$

The conservation of energy, expressed by Eq. (3.10), then yields the following relationship between the above two momentum quantities:

$$\vec{p}_T^{\,2} + p_z^2 = 2em\Phi_a\,. \tag{3.28}$$

The same relationship also follows from Eqs. (3.27) directly, but those equations were already formed by using the conservation of the Hamiltonian $\mathscr{H}$ along the particle trajectory. To summarize the above, we have found that

$$\vec{\rho}\,'(z) = \vec{p}_T/p_z\,, \qquad p_z = \sqrt{2em\Phi_a - p_T^2}\,, \tag{3.29}$$

and these equations lead directly to,

$$\vec{\rho}\,''(z) = \frac{\vec{p}\,'_T}{p_z} - \frac{\vec{p}_T}{p_z^2}\,p'_z = \frac{\vec{p}\,'_T}{p_z} - \frac{\vec{p}_T}{p_z^3}\,(em\Phi'_a - p_T p'_T)$$
$$= \frac{\vec{p}\,'_T}{p_z} - \frac{\vec{p}_T}{p_z^3}\,(emE_z - p_T p'_T) \rightarrow$$

$$\vec{\rho}\,''(z) = \frac{1}{p_z}\left(\vec{p}_T - \frac{p_T p'_T}{p_z^2}\,\vec{p}_T\right) + emE_z\,\frac{\vec{p}_T}{p_z^3}\,. \tag{3.30}$$

It can be seen that the last term of Eq. (3.30) cancels the E_z term of Eq. (3.19). Substitution of $\vec{\rho}\,'(z)$ and $\vec{\rho}\,''(z)$ from Eqs. (3.29) and (3.30) into Eq. (3.19) then yields two coupled first-order differential equations in p_x and p_y. Once these equations have been solved, one needs only to perform the quadrature of $\vec{\rho}\,'(z) = \vec{p}_T/p_z$ to obtain the spatial coordinate $\vec{\rho}(z)$. However, instead of explicitly giving the differential equations for p_x and p_y here as a replacement of Eqs. (3.18) or (3.19), we shall first generalize to relativistic equations of motion. The reader who wishes to skip this development can proceed directly to Eq. (3.62).

3.3 RELATIVISTIC EQUATIONS OF MOTION

Since the epochal publication of Albert Einstein in 1905 on the special theory of relativity, it has been clear that "Newtonian" mechanics is only an approximation that becomes inaccurate when velocities v appear that are not small with respect to the velocity of light in a vacuum: $c \approx 2.998 \times 10^8$ ms/s. However, the laws of Newton do retain the same form when a transformation from one frame of reference with coordinate system x, y, z is made to another frame of reference with coordinate system x_1, y_1, z_1 which moves with a uniform velocity with respect to the first. That retention of form (we say these laws are *invariant* with respect to *Galilean* transformations) makes it impossible to decide from the *form* of Newton's laws in which frame of reference an observer is. Frames of reference with uniform velocity are called *inertial*. As far as the mechanical laws of physics are concerned, all inertial frames are on the same footing. The mechanical laws of nature are the same everywhere in the universe if we ignore frames of reference that have accelerations compared to inertial frames. For example, a rotating frame of reference such as the Earth with respect to its short-term motion along its orbit (which can be considered approximately linear and uniform in velocity), introduces an apparent force in east or west direction for any motion north or southwards: the *coriolis* force which must be added explicitly to the kinematic equations in the rest frame of the rotating Earth. However, the Galilean transformations result in the ordinary vector addition of velocities $\vec{v}_1 + \vec{v}_2$ when we compare Frame C to Frame A, given that Frame B moves with velocity $\vec{v}_1$ with respect to Frame A, and Frame C moves with velocity $\vec{v}_2$ with respect to Frame B.

The relativistic complications occur when electromagnetic forces are also taken into account. For example, the $\vec{E}$ and $\vec{B}$ fields can be determined from potentials Φ and $\vec{A}$ (Jackson 1975). The scalar potential Φ obeys the Laplace equation in the absence of charge sources (Corbon and Stehle 1960):

$$\Delta\Phi - \frac{1}{c^2}\frac{\partial^2\Phi}{\partial t^2} = 0\,. \tag{3.31}$$

Under the Galilean transformation $x_1 = x$, $y_1 = y$, $z_1 = z - vt$, $t_1 = t$, this equation becomes

$$\Delta_1\Phi - \frac{v^2}{c^2}\frac{\partial^2\Phi}{\partial z_1^2} + \frac{2v}{c^2}\frac{\partial^2\Phi}{\partial z_1\,\partial t_1} - \frac{1}{c^2}\frac{\partial^2\Phi}{\partial t^2} = 0\,, \tag{3.32}$$

i.e., the equation has indeed changed its form! Consequently, there must be some zero-velocity frame in which the simplest form (3.31) holds, and this frame would then have some sort of absolute meaning. In both Eqs. (3.31) and (3.32), c is the velocity of light with respect to this frame. We conclude that Eq. (3.32) pertains to a frame with a nonzero velocity because it has a different form.

This state of affairs led researchers around the turn of the century to search for an absolute frame of reference, the "ether." Not only did the famous experiment of Michelson and Morley (Bergmann 1976) fail to show the existence of an absolute frame, Einstein reasoned that it was much more satisfactory to assume that the velocity of light in a vacuum is the same in all frames of reference moving at constant velocities with respect to each other. To maintain this simple hypothesis, he emphasized a different class of coordinate transformations now known as *Lorentz transformations*. All Lorentz transformations leave invariant the four-dimensional length ds (not to be confused with the ds of previous sections) upon transformation from one inertial frame to another. The new length ds is given by

$$ds^2 = c^2\,dt^2 - dx^2 - dy^2 - dz^2\,. \tag{3.33}$$

One of the simplest Lorentz transformations is for frames with relative velocity $\vec{v} = v\hat{z}$:

$$x_1 = x\,, \qquad y_1 = y\,, \qquad z_1 = \gamma(z - vt)\,, \qquad t_1 = \gamma\left(t - \frac{vz}{c^2}\right), \tag{3.34}$$

where $\gamma^2 = 1/[1 - (v^2/c^2)]$ is a constant factor determined by the relative velocity v. It is common practice to refer to the ratio v/c as β, i.e., $\beta \equiv v/c$. An important consequence of Eq. (3.34) is the addition of velocities: if Frame C moves at velocity v_2 in Frame B, and Frame B at velocity v_1 in Frame A (all in the z direction), then Frame C moves at v in Frame A with

$$v = \frac{v_1 + v_2}{1 + (v_1 v_2/c^2)}\,. \tag{3.35}$$

This summed velocity cannot exceed c. When either v_1 or v_2 equals c, we find that $v = c$, no matter what the other velocity is. The summed velocity of Eq. (3.35) reduces approximately to the ordinary linear sum when $v_1 v_2 \ll c^2$.

The special theory of relativity thus states that we must cast the equations of electron optics, which mix mechanical and electromagnetic quantities in the trajectory equations, into a form that remains invariant under Lorentz transformations. One sure-fire way of achieving this is to reorganize all quantities into *four-vectors*. In the same sense that the spatial coordinates x, y, z transform into each other under Galilean transformations, the components of a four-vector will transform into each other under a Lorentz transformation. The following sets form four-vectors:

$$\begin{aligned}
&\vec{\mathbb{X}} = (x, y, z, ict) && \text{space, time coordinates},\\
&\vec{\mathbb{P}} = \gamma(m_0 v_x, m_0 v_y, m_0 v_z, im_0 c) && \text{momentum, energy}, \qquad (3.36)\\
&\vec{\mathbb{A}} = (A_x, A_y, A_z, i\Phi/c) && \text{vector, scalar potentials}.
\end{aligned}$$

The quantity $\mathscr{E}_k = m_0 c^2(\gamma - 1)$ is the relativistic equivalent of the kinetic energy T. When $v \ll c$, we may approximate $\gamma \approx 1 + (v^2/2c^2)$ to obtain the nonrelativistic value $\mathscr{E}_k \approx m_0 v^2/2 = T$.

The Maxwell equations can be formulated in terms of the potentials Φ and $\vec{A}$, which each obey d'Alembertian equations of the form $\Delta\Phi - c^{-2}\,\partial^2\Phi/\partial t^2 = 0$ in source-free regions, provided the potentials are related by the Lorentz condition $\vec{\nabla} \cdot \vec{A} + \mu\epsilon\,\partial\Phi/\partial t = 0$. The corresponding $\vec{E}$ and $\vec{B}$ fields are then given by

$$\vec{E} = -\vec{\nabla}\Phi - \frac{\partial \vec{A}}{\partial t}, \qquad \vec{B} = \vec{\nabla} \times \vec{A}. \tag{3.37}$$

In the special theory of relativity, the six quantities $E_x, E_y, E_z, B_x, B_y, B_z$ are replaced by the nonzero off-diagonal components $\mathbb{F}_{ij}$ of an antisymmetric tensor $\vec{\vec{\mathbb{F}}}$ defined by

$$\mathbb{F}_{ij} \equiv \frac{\partial \mathbb{A}_j}{\partial \mathbb{X}_i} - \frac{\partial \mathbb{A}_i}{\partial \mathbb{X}_j}, \quad 1 \leq ij \leq 4. \tag{3.38}$$

Clearly all f_{ii} are zero, and $\mathbb{X}_i$, $\mathbb{A}_j$, etc., are chosen from the components of the four-vectors $\vec{\mathbb{X}}$ and $\vec{\mathbb{A}}$ given in Eq. (3.36). From the preceding three equations, one can then find,

$$\vec{\vec{\mathbb{F}}} = \begin{bmatrix} 0 & B_z & -B_y & -\dfrac{iE_x}{c} \\ -B_z & 0 & B_x & -\dfrac{iE_y}{c} \\ B_y & -B_x & 0 & -\dfrac{iE_z}{c} \\ \dfrac{iE_x}{c} & \dfrac{iE_y}{c} & \dfrac{iE_z}{c} & 0 \end{bmatrix}. \tag{3.39}$$

The Lorentz transformation (3.34) can then be described by the tensor

$$\vec{\vec{\mathbb{S}}} = \begin{bmatrix} 1 & 0 & 0 & 0 \\ 0 & 1 & 0 & 0 \\ 0 & 0 & \gamma & i\beta\gamma \\ 0 & 0 & -i\beta\gamma & \gamma \end{bmatrix} \tag{3.40}$$

with $\beta \equiv v/c$. The reader is reminded that $\vec{\vec{\mathbb{F}}}$ is a tensor representation of a four-dimensional curl written symbolically as $(\partial/\partial\vec{\mathbb{X}}) \times \vec{\mathbb{A}}$. Often, a distinction is made between *contravariant transformation* (i.e., as $\vec{\mathbb{X}}$), and *covariant transformation* (i.e., as $\partial/\partial\vec{\mathbb{X}}$), but it is not needed here. Consequently, this tensor will transform into

$$\vec{\vec{\mathbb{F}}}_L = \vec{\vec{\mathbb{S}}} \cdot \vec{\vec{\mathbb{F}}} \cdot \vec{\vec{\mathbb{S}}}^T , \tag{3.41}$$

where $\vec{\vec{\mathbb{S}}}^T$ is the transpose of tensor $\vec{\vec{\mathbb{S}}}$ (and is equal to the inverse of $\vec{\vec{\mathbb{S}}}$ as this tensor is unitary). If we write out the six nonzero components of $\vec{\vec{\mathbb{F}}}_L$ explicitly, we find from Eq. (3.41),

$$\begin{aligned} \mathbb{F}_{23} &= B_{1x} = \gamma\left(B_x + \frac{vE_y}{c^2}\right), & \mathbb{F}_{14} &= -\frac{iE_{1x}}{c} = -i\gamma\left(\frac{E_x}{c} - \beta B_y\right), \\ \mathbb{F}_{31} &= B_{1y} = \gamma\left(B_y - \frac{vE_x}{c^2}\right), & \mathbb{F}_{24} &= -\frac{iE_{1y}}{c} = -i\gamma\left(\frac{E_y}{c} + \beta B_x\right), \\ \mathbb{F}_{12} &= B_{1z} = B_z , & \mathbb{F}_{34} &= -\frac{iE_{1z}}{c} = -\frac{iE_z}{c} . \end{aligned} \tag{3.42}$$

The fields in the direction of v (namely E_z and B_z) are unchanged upon transformation, but the transverse vectors $\vec{E}_T$ and $\vec{B}_T$ transform explicitly, according to Eq. (3.42), as

$$\vec{E}_T \rightarrow \gamma(\vec{E} + \vec{v} \times \vec{B})_T , \qquad \vec{B}_T \rightarrow \gamma[\vec{B} - (\vec{v} \times \vec{E})/c^2]_T . \tag{3.43}$$

The nonrelativistic equation of force is now replaced by

$$\frac{d}{dt}(m\gamma\dot{\vec{r}}) = -e(\vec{E} + \dot{\vec{r}} \times \vec{B}) , \tag{3.44}$$

but in this form it is not expressed in terms of relativistically invariant quantities. If, however, we replace dt by the "proper-time" interval $d\tau = \gamma^{-1}\, dt$ then Eq. (3.44) can easily be shown to be of the form

$$\frac{d\mathbb{P}_i}{d\tau} = -\frac{e}{m_0} \sum_{j=1}^{4} \mathbb{F}_{ij}\mathbb{P}_j\,, \quad i = 1, 2, 3\,, \tag{3.45}$$

i.e., the equation has been written as the *spatial* part of a relativistically invariant equation. Equation (3.44) is in itself not incorrect; it is simply not relativistically invariant with respect to Lorentz transformations of frames of reference (which differ by a constant velocity $\vec{v}$). Equation (3.45) needs to be cast into a more useful form, and to do so we will write the four-vector $\vec{\mathbb{P}}$ as the (3 + 1)-vector ($\vec{p}$, p_4) with $\vec{p} = \gamma m_0 \dot{\vec{r}}$ and $p_4 = i\gamma m_0 c$. That results in

$$\frac{dp_i}{d\tau} = -\frac{e}{m_0}\left(\sum_{j=1}^{3} \mathbb{F}_{ij}p_j + \mathbb{F}_{i4}p_4\right), \quad i = 1, 2, 3\,, \tag{3.46}$$

as a replacement for Eq. (3.45), but in fact we need this last equation only for the *transverse* components p_x ($i = 1$) and p_y ($i = 2$) because

$$p^2 = \gamma^2 m_0^2 \dot{\vec{r}}^2 = \gamma^2 m_0^2 \beta^2 c^2 = m_0^2 c^2(\gamma^2 - 1)\,, \tag{3.47}$$

which establishes the following relationship between p_z ($i = 3$) and the two transverse components:

$$p_z = m_0 c\sqrt{(\gamma^2 - 1) - \frac{p_T^2}{m_0^2 c^2}}\,. \tag{3.48}$$

On the other hand, the left-hand side of Eq. (3.46) can be rewritten as,

$$\frac{dp_i}{d\tau} = \gamma\,\frac{dp_i}{dt} = \gamma \dot{z} p_i' = \frac{p_z p_i'}{m_0}\,. \tag{3.49}$$

Hence, Eq. (3.46) becomes

$$p_i' = -\frac{e}{p_z}\left(\sum_{j=1}^{3} \mathbb{F}_{ij}p_j + \mathbb{F}_{i4}p_4\right), \quad i = 1, 2\,. \tag{3.50}$$

The relativistic parameter γ is related to the relativistic kinetic energy $\mathscr{E}_k = m_0 c^2(\gamma - 1)$, and with Eq. (3.48) we obtain

$$\gamma = 1 + \frac{\mathscr{E}_k}{m_0 c^2}\,, \qquad p_z = \sqrt{2m_0\mathscr{E}_k\left(1 + \frac{\mathscr{E}_k}{2m_0 c^2}\right) - p_T^2}\,. \tag{3.51}$$

Thus, Eq. (3.50) with supporting Eqs. (3.51) replace the nonrelativistic equations of motion in a momentum representation of the transverse spatial part of a four-vector. Referring to the definition of $\vec{\vec{\mathbb{F}}}$ in Eq. (3.39), we observe that the first three terms in Eq. (3.50) pertain to the $\vec{v} \times \vec{B}$ term, and the last one to the $\vec{E}$ term of the total electric + magnetic Lorentz force.

Let us now replace the $\mathbb{F}_{ij}$ factors in Eq. (3.50) by the $\vec{E}$ and $\vec{B}$ fields explicitly, even though these are not relativistically invariant quantities. The reason for doing so is that we will be applying trajectory equations to a situation in which the electromagnetic fields are at rest with respect to the coordinate system of interest whereas the particles are moving. Accordingly, we can dispense with more complicated relativistically invariant expressions for fields as long as we are more careful about quantities such as mass, momentum, and energy of the particles. The replacement yields

$$\vec{p}\,'_T = -\frac{e}{p_z}\,(m_0\gamma\vec{E} + \vec{p} \times \vec{B})_T\,, \tag{3.52}$$

for the transverse components. To complete our quest for useful equations we finally define an equivalent new kinetic energy U_k with

$$U_k \equiv \frac{p^2}{2m_0}\,. \tag{3.53}$$

If we apply Eq. (3.47) for p^2 and then replace γ by the expression for it in Eq. (3.51), we find that

$$U_k = \mathscr{E}_k\left(1 + \frac{\mathscr{E}_k}{2m_0c^2}\right), \tag{3.54}$$

i.e., that U_k reduces to the nonrelativistic kinetic energy T at velocities $v \ll c$. From the preceding formulas and definitions, it then follows than the transverse gradient of U_k is given by

$$\vec{\nabla}_T U_k = \vec{\nabla}_T \mathscr{E}_k + \frac{\mathscr{E}_k}{m_0c^2}\,\vec{\nabla}_T \mathscr{E}_k = \left(1 + \frac{\mathscr{E}_k}{m_0c^2}\right)\vec{\nabla}_T \mathscr{E}_k = \gamma\vec{\nabla}_T \mathscr{E}_k\,. \tag{3.55}$$

It is still true in the framework of the special theory of relativity that the Hamiltonian is conserved along a trajectory, i.e., that $d\mathscr{H}/ds = 0$, but the Hamiltonian is now defined more generally as

$$\mathscr{H}(s) \equiv \mathscr{E}_k(s) - e\Phi(s)\,. \tag{3.56}$$

This expression holds in the rest frame of electromagnetic fields, and conservation of $\mathcal{H}(s)$ along s yields

$$\vec{\nabla}_T \mathscr{E}_k = e\vec{\nabla}_T \Phi_a = -e\vec{E} \ . \tag{3.57}$$

We have made use of the fact that $\vec{\nabla}\Phi = \vec{\nabla}\Phi_a$ here. Use this expression in Eq. (3.55) to obtain $\vec{\nabla}_T U_k = -e\gamma\vec{E}$, and then insert that into (3.52) to obtain

$$\vec{p}\,'_T = -\frac{m_0}{p_z}\left[\vec{\nabla}U_k - \frac{e}{m_0}(\vec{p}\times\vec{B})\right]_T , \qquad p_z = \sqrt{2m_0U_k - p_T^2} ,$$
$$U_k = \mathscr{E}_k\left(1 + \frac{\mathscr{E}_k}{2m_0c^2}\right) . \tag{3.58}$$

These are useful equations for trajectories in situations where the fields are static, but for which relativistic effects due to high particle energies must be taken into account. For the sake of making the equations somewhat easier to convert to nonrelativistic situations, and also to prepare for numerical calculations, it is advantageous to replace the relativistic momentum vector $\vec{p}$ by a new quantity $\vec{P}$. So far, we have not distinguished in notation between relativistic $\vec{p}$ and nonrelativistic $\vec{p}$ because the former transforms into the latter as parameter $\beta \equiv v/c \rightarrow 0$. The new relativistic "momentum" is

$$\vec{P} \equiv \frac{\vec{p}}{\sqrt{(2m_0)}} , \tag{3.59}$$

which has the dimension of $(\text{energy})^{1/2}$, and is therefore conveniently measured in units $\sqrt{\text{eV}}$ (where eV is an abbreviation of *electron-volt*; a very convenient unit of energy in electron optics). With this conversion, the trajectory equations (3.58) become,

$$\vec{P}\,'_T = -\frac{1}{2P_z}\left[\vec{\nabla}U_k - \sqrt{\frac{2}{m_0}}\, e(\vec{P}\times\vec{B})\right]_T , \qquad P_z = \sqrt{U_k - P_T^2} ,$$
$$U_k = \mathscr{E}_k\left(1 + \frac{\mathscr{E}_k}{2m_0c^2}\right) . \tag{3.60}$$

These equations are more easily understood by remembering that $\mathscr{E}_k = e\Phi_a$ (see Eq. (3.56), and invoke the conservation of $\mathcal{H}$ in analogy to Eq. (3.9)). We can obtain the three components P_x, P_y, and P_z at all points s of a trajectory by one quadrature of Eqs. (3.60) if Φ_a and $\vec{B}$ are known everywhere. The second quadrature is obtained from the direction tangents at s, namely,

$$x'(z) = \frac{P_x(s)}{P_z(s)} , \qquad y'(z) = \frac{P_y(s)}{P_z(s)} . \tag{3.61}$$

3.4 FURTHER DISCUSSION OF THE EQUATIONS OF MOTION

The equations of motion have appeared in quite a number of forms in the preceding sections, and it seems useful to review these before moving on to the topic of solving these equations. The form we wish to use is mostly dictated by the situation at hand; the equations for an electron mirror, for example, will be described better in terms of an independent path parameter $s = vt$ than in terms of a preferred Cartesian coordinate z.

The most general equations developed so far are valid in the framework of the special theory of relativity: Eqs. (3.50) with (3.51) hold for a frame in which the electromagnetic fields are at rest and in which the momentum $\vec{p}$ is relativistic as given in Eq. (3.36). While it is so that these equations have been formulated in terms of a Cartesian coordinate z as independent variable, it is not difficult to generalize to a path coordinate s, or, relativistically, to τ or even $c\tau$, but that generalization is implied by Eq. (3.45) and it is easily adapted by procedures that are very similar to those discussed in Section 3.2.

Because most situations in electron optics can be handled in the rest frame of the electromagnetic fields $\vec{E}$ and $\vec{B}$, it is usually not necessary to invoke the electromagnetic field tensor $\vec{\vec{\mathbb{F}}}$ of Eq. (3.39) in the equations of motion. As a result, Eqs. (3.58) are obtained in terms of momenta, and Eqs. (3.60) in terms of equivalent square-root energies.

This last-mentioned form of the equations of motion is most useful in describing the transition from relativistic to nonrelativistic equations. For this purpose we need two transitions in Eq. (3.60):

(i) The relativistic momentum vector $\vec{p}$ must make a transition into the nonrelativistic equivalent, $\vec{p} = m\dot{\vec{r}}$, i.e., $\gamma \rightarrow 1$.

(ii) The relativistic kinetic energy U_k must devolve into its nonrelativistic equivalent, $U_k \rightarrow \mathscr{E}_k$.

This occurs up to terms of order β^2 by ignoring terms of $O(\mathscr{E}_k/m_0c^2)$ compared to unity. If we apply $\mathscr{E}_k = e\Phi_a$ (as explained in connection with Eq. (3.60)), the following nonrelativistic equations will result:

$$\vec{p}_T' = \frac{e}{2P_z}\left[\vec{E}_T + \sqrt{\frac{2}{m_0}}\,(\vec{P} \times \vec{B})_T\right], \qquad P_z = \sqrt{e\Phi_a - P_T^2}\,. \tag{3.62}$$

These equations also follow directly from the development outlined at the end of Section 3.2.

Clearly, Eqs. (3.60), when integrated to yield $\vec{P}_T(z)$, lead to the correct trajectory coordinates $\vec{\rho}(z)$ via relativistic quadrature of Eqs. (3.61), and nonrelativistically upon replacing U_k by $\mathscr{E}_k$. Hence with respect to *trajectory* equations, the relativistic correction to the nonrelativistic equations (3.62) is simply the replacement of $\mathscr{E}_k$ by U_k, i.e., by the replacement

$$e\Phi_a \rightarrow e\Phi_a\left(1 + \frac{e\Phi_a}{2m_0c^2}\right). \tag{3.63}$$

In terms of the equivalent *refractive index* of Eq. (2.29), the transition to relativistic velocities can be made by choosing

$$n(s) = \kappa\sqrt{\Phi_a(s)}\left[1 + \frac{e\Phi_a(s)}{2m_0c^2}\right]^{1/2}, \tag{3.64}$$

in the otherwise classical expressions. The neglect of relativistic effects amounts to the neglect of the correction term $e\Phi_a/2m_0c^2$ with respect to unity in the refractive index. As $2m_0c^2 \approx 1022$ keV, it follows that an error of at most 1.5% in the refractive index is made in neglecting this term when the highest potential in a situation is 30 kV. The factor 1.5% is obtained from the ratio of 30 to 1022, multiplied by the power 1/2 for the first term of the binomial expansion of the term in square brackets in Eq. (3.64). This is a useful guideline for the electron optics designer of cathode-ray tubes, as 30 kV is usually a typical high voltage.

3.5 TWO-DIMENSIONAL GEOMETRIES

Thus far, we have discussed the general case in which electron trajectories are not restricted in three-dimensional space by the fields to a plane. In practice, many situations occur in which only two spatial coordinates are needed to describe the trajectories. This section is devoted to the two most common ones.

3.5.1 Cylindrical (Axial) Symmetry

Whenever the guiding fields are purely electrostatic and the potential is axisymmetric (i.e., possesses cylindrical symmetry), it follows from any of the preceding equations of motion that electron trajectories lie in a *meridional* plane. That is, electrons continue to move in a fixed plane defined by the axis of cylindrical symmetry and the initial velocity. The equations of motion contain only functions of, and derivatives with respect to, the coordinates ρ and z, where $\rho^2 = x^2 + y^2$. Consider Eqs. (3.62) as an example; these now become

$$P'_\rho = -\frac{e}{2P_z} E_\rho , \qquad E_\rho = -\partial \Phi_a / \partial \rho , \qquad P_z = \sqrt{e\Phi_a - P_\rho^2} , \tag{3.65}$$

where the accelerating potential $\Phi_a(r, z)$ is a function of the scalar coordinates ρ and z. As there is no P_ϕ component, the initial azimuthal angle ϕ of a meridional electron obeying these equations is maintained, and the electron trajectory lies entirely in a fixed plane through the z axis.

The axial symmetry of $\Phi_a(\rho, z)$, hence of $\Phi(\rho, z)$, allows us to express it for nonzero ρ as a power series in ρ^2 with coefficients given by the derivatives of the axial potential $\Phi_0(z) \equiv \Phi(0, z)$ with respect to z. This is demonstrated most elegantly as follows. Let $\vec{\nabla}_T$ be the transverse gradient operator as before, i.e.,

$$\vec{\nabla}_{0T} \equiv \hat{x} \frac{\partial}{\partial x_0} + \hat{y} \frac{\partial}{\partial y_0} . \tag{3.66}$$

in which expression the subindex "0" is added to indicate derivatives with respect to x_0 and y_0. The Taylor-series expansion of $\Phi(\vec{\rho}, z)$, even in the nonaxisymmetric case, around the z axis can then be written formally as

$$\Phi(\vec{\rho}, z) = [\exp(-\vec{\rho} \cdot \vec{\nabla}_{0T}) \Phi(\vec{\rho}_0, z)]_{\vec{\rho}_0 = 0} , \tag{3.67}$$

in the understanding that the exponential operator $\exp(-\vec{\rho} \cdot \vec{\nabla}_{0T})$ is merely a convenient concatenation of the Taylor series. For axial symmetry, Eq. (3.67) can be written as an average over all possible azimuthal angles,

$$\Phi(\rho, z) = \frac{1}{2\pi} \int_0^{2\pi} d\phi [\exp(|\vec{\rho}| \, |\vec{\nabla}_{0T}| \cos \phi) \Phi(\vec{\rho}_0, z)]_{\vec{\rho}_0 = 0} . \tag{3.68}$$

Continuing the convention of representing power series of differential operators by formal "functions", we note that Eq. (3.68) is the integral representation of

$$\Phi(\rho, z) = [I_0(\rho|\vec{\nabla}_{0T}|)\Phi(\vec{\rho}_0, z)]_{\vec{\rho}_0=0}$$
$$= \sum_{m=0}^{\infty}\left[\frac{1}{(m!)^2}\left(\frac{\rho^2\Delta_0}{4}\right)^m \Phi(\vec{\rho}_0, z)\right]_{\vec{\rho}_0=0}, \qquad (3.69)$$

where $I_0(\ldots)$ is a modified Bessel function of order zero that is finite for zero argument (Abramowitz and Stegun 1964). Finally, we apply Laplace's equation for the electrostatic potential Φ to replace Δ_0 by $\partial^2/\partial z^2$ in Eq. (3.69), and then one may set $\vec{\rho}_0 = 0$ without further ado to obtain

$$\Phi(\rho, z) = \sum_{m=0}^{\infty} \frac{1}{(m!)^2}\left(-\frac{\rho^2}{4}\frac{\partial^2}{\partial z^2}\right)^m \Phi(0, z)\,. \qquad (3.70a)$$

This is the desired expression, and it can be considered as the formal expansion of the operator equation,

$$\Phi(\rho, z) = J_0\left(\rho\frac{\partial}{\partial z}\right)\Phi(0, z)\,, \qquad (3.70b)$$

where $J_0(\ldots)$ is the ordinary Bessel function of the first kind of order zero and finite at zero argument. One will find three or four of the expansion terms of this equation derived directly in a somewhat elaborate fashion in many electron optics texts. Here we immediately obtain from this formal expression,

$$\Phi(\rho, z) = \Phi_0 - \tfrac{1}{4}\Phi_0''\rho^2 + \tfrac{1}{64}\Phi_0^{(4)}\rho^4 - \tfrac{1}{2304}\Phi_0^{(6)}\rho^6 + \cdots\,. \qquad (3.71)$$

Note here that $\Phi_0^{(n)}$ is the nth partial derivative of the axial potential $\Phi(0, z)$ with respect to z. In a later section we will observe that the first two terms of Eq. (3.71) suffice for the *paraxial* or *Gaussian* approximation of the equations of motion, which gives rise to perfect focussing of a beam of electrons that enter a lensing region all parallel to the z axis and all with one velocity. A lensing region is defined by $\partial\Phi(\rho, z)/\partial\rho \neq 0$ (provided the mutual repulsion of the electrons is ignored). The higher-order terms in Eq. (3.71) give rise to *spherical aberrations* (see Chapter 8).

3.5.2 Planar Symmetry

Another situation in which two-dimensional equations of motion suffice occurs when the guiding electrostatic potential $\Phi(x, z)$ is independent of one Cartesian coordinate, say y, and $\Phi(-x, z) = \Phi(x, z)$. The velocity component of an electron in the y direction will be unchanged; the electron *drifts* in the y direction. Equations (3.62), for example, will then become

$$P'_x = -\frac{e}{2P_z} E_x \quad \text{with} \quad E_x = -\frac{\partial \Phi_a}{\partial x}, \qquad P_z = \sqrt{e\Phi_a - P_x^2}\,. \tag{3.72}$$

It is sufficient to examine the projection of the electron trajectory upon a y = constant plane; however, one should bear in mind that electron densities observed at an endpoint s_f or z_f do depend upon the drift velocities in the y direction! Instead of Eq. (3.67), we now have

$$\Phi(x, z) = \left[\exp\left(\pm x \frac{\partial}{\partial x_0}\right)\Phi(x_0, z)\right]_{x_0=0}, \tag{3.73}$$

or, in complete analogy to the steps outlined in Eqs. (3.68)–(3.71),

$$\begin{aligned}\Phi(x, z) &= \frac{1}{2}\left\{\left[\exp\left(x \frac{\partial}{\partial x_0}\right) + \exp\left(-x \frac{\partial}{\partial x_0}\right)\right]\Phi(x_0, z)\right\}_{x_0=0} \\ &= \left[\cosh\left(x \frac{\partial}{\partial x_0}\right)\Phi(x_0, z)\right]_{x_0=0} \\ &= \sum_{m=0}^{\infty}\left[\frac{1}{(2m)!}\left(x^2 \frac{\partial^2}{\partial x_0^2}\right)^m \Phi(x_0, z)\right]_{x_0=0} \\ &= \sum_{m=0}^{\infty}\left[\frac{(-1)^m}{(2m)!}\left(x^2 \frac{\partial^2}{\partial z^2}\right)^m\right]\Phi(0, z)\,,\end{aligned} \tag{3.74}$$

so that the symbolic form is obtained:

$$\Phi(x, z) = \cos\left(x \frac{\partial}{\partial z}\right)\Phi(0, z)\,, \tag{3.75}$$

which is equivalent to the last part of Eq. (3.74), of which the first few terms are,

$$\Phi(x, z) = \Phi_0 - \tfrac{1}{2}\Phi_0'' x^2 + \tfrac{1}{24}\Phi_0^{(4)} x^4 - \tfrac{1}{720}\Phi_0^{(6)} x^6 + \cdots. \tag{3.76}$$

3.6 INTEGRATION OF THE EQUATIONS OF MOTION

In the preceding sections we have encountered equations of motion in various forms, all of which are coupled first-order or second-order ordinary differential equations (ODEs) with initial conditions. In Eqs. (3.18) or (3.19) the equations are formulated in terms of the two transverse coordinates for coordinates for nonrelativistic velocities. In Eq. (3.26) the independent variable is the path parameter s, rather than a preferred Cartesian coordinate z. In Eq. (3.62), together with (3.61), a momentum formulation is utilized so that several first-order ODEs replace the two second-oder ODEs. Equations (3.58) or (3.60) represent forms of the equations of motion that allow for relativistic velocities. If, however, entirely invariant forms of the relativistic equations are required, then these last-mentioned equations must be replaced by Eqs. (3.50) with (3.51), or, even more generally, by Eqs. (3.45).

Whatever set of trajectory equations are selected for use, one can seldom obtain analytical solutions. The guiding electromagnetic fields are nearly always too complicated. Even when analytical expressions for $\vec{E}$ and $\vec{B}$ are available, it is well-nigh impossible to extract analytical solutions of the equations of motion except for the simplest situations. Certain canonical situations such as Gaussian optics of electron lenses are discussed later, but here it is stressed that *numerical* solution of the equations of motion is almost always the only tractable approach that will yield acceptable accuracy. Numerical solution these days implies solution with the aid of an electronic digital computer. We will not discuss older, and now quite outdated, analog methods. Chief among these were the simulation of electrostatic potential fields by means of a *rubber sheet* or an *electrolytic tank* (Grivet 1972).

Because there are many good texts on numerical solution of ODEs we will simply outline the major features of useful procedures. Let us take the equations of motion in a purely electrostatic field in a two-dimensional planar geometry as an example:

$$x'' = -\frac{1 + x'^2}{2\Phi_a} (E_x - x'E_z) ,$$

or (3.77)

$$P'_x = -\frac{e}{2P_z} E_x \quad \text{with} \quad P_z = \sqrt{e\Phi_a - P_x^2} .$$

Here, all field quantities are functions of z in principle. It is not important for this discussion that Eq. (3.77) is a special case because the same procedures hold for more general three-dimensional equations, but the extra complications of those detract from the essential features which are more easily understood from what follows.

All higher-order ODEs (as in the x'' equation above) can be converted into a set of first-order ODEs, possibly coupled. For example, set

$x'(z) \equiv p(z)$ (do not confuse with momentum), and then the first of Eqs. (3.77) becomes

$$p'(z) = -\frac{1+p^2}{2\Phi_a}\,[E_x(z) - pE_z(z)]\,, \qquad x'(z) = p(z)\,. \qquad (3.78)$$

In this example, we first solve the first ODE in Eq. (3.78) to obtain $p(z)$ as a function of z, and then solve the second ODE which results from substitution into the right-hand side. An initial condition is needed for both equations, i.e., one needs to know $p(z_0)$ and $x(z_0)$ to obtain unique solutions.

Actually, the momentum forms in Eq. (3.77) are equivalent to Eq. (3.78) because $x'(z) \equiv P_x(z)/P_z(z)$ provides the additional first-order ODE needed to generate $x(z)$ from the two components of these momenta. Let us therefore work further with the momentum forms in Eq. (3.77). The crudest approach, which suffices to illustrate what is at hand, is *Euler's method*, in which we assume that $E_x(z)$ is known and that $P_x(z)$ and $P_z(z)$ have already been found (Bajpai et al. 1977). A simple Taylor expansion yields

$$P_x(z+\Delta z) = P_x(z) + \Delta z\, P_x'(z) + O(\Delta z^2)\,. \qquad (3.79)$$

Insert Eq. (3.77) for $P_x'(z)$ into the second term on the right-hand side of this equation to obtain an estimate of $P_x(z+\Delta z)$ in terms of the already known or obtained quantities. If $P_x(z_0)$ and $P_z(z_0)$ are known for an initial position z_0, then this procedure will generate $P(z)$ as a function of z at intervals Δz. The Euler procedure is somewhat crude because the error is $O(\Delta z^2)$ and consequently quite small intervals Δz are needed to avoid unacceptable cumulative errors. Euler's method is a so-called *one-step* method in which (Chapra and Canale 1985)

$$(\text{new value}) = (\text{old value}) + (\text{slope}) \times (\text{step size})\,. \qquad (3.80)$$

A number of modifications of one-step methods are possible which increase the accuracy of the Δz steps. One of the most used of such modifications is the class of *Runge-Kutta* methods which estimate an average slope of the $P_x(z)$ function over Δz from a number of estimated intermediate values $P_x(z + \alpha_i\, \Delta z)$, in which $1 \leq i \leq n$ for an nth-order Runge-Kutta method, and $\alpha_i < 1$.

Multistep methods use information not from one previous point but from two or more already calculated values $P_x(z)$, $P_x(z - \Delta z), \ldots$, to estimate $P_x(z+\Delta z)$. These methods can provide greater accuracy in larger Δz steps by giving information about the *curvature* of $P_x(z)$ at z.

The price to be paid is that the resulting difference equations are somewhat more complicated than (3.79) and its one-step modifications.

Common to all methods is a discretization process in which a trajectory is broken down (discretized) into Δz steps. When we need to solve a second-order ODE (or set of such), then improved accuracy can be obtained (Hockney and Eastwood 1981) by means of a "leapfrog" scheme: interval n to $n+1$ for $P_x(z)$ but interval $n-\frac{1}{2}$ to $n+\frac{1}{2}$ for $x(z)$ after solving for $P_x(z)$.

Yet another problem in electron optics trajectory equations is the possible dependence of the electromagnetic fields upon the very particle beams they influence. If, for example, Eqs. (3.77) are applied to a beam of electrons, then the equation for one electron will include in the $\vec{E}$ field the Coulomb (space-charge) forces of all other electrons in the beam. Those forces are functions of the locations of all N electrons in the beam; worse than that, the equation for $x_i(z)$ of the ith electron will also depend upon $x_j(z_1)$ at all other locations z_1. The N equations are highly coupled and nonlinear, and a straightforward approach via "marching" methods (such as Euler or Runge-Kutta) is not possible. Iterative techniques (Hockney and Eastwood 1981) are usually applied in such cases: a profile of $N-1$ electrons is assumed to give rise to a space-charge force on the Nth electron. The trajectory of this electron is then calculated as before. The same is repeated for all N electrons in turn. Then the result gives rise to a new profile of N electrons, which in turn is used to recalculate the trajectories. The iterative procedure stops when the new trajectories of the $(n+1)$st iteration differ insignificantly from those of the nth (Kirstein et al. 1967).

REFERENCES

Abramowitz, M. and I.A. Stegun (1964) *Handbook of Mathematical Functions*, National Bureau of Standards *Appl. Math. Series* **55**, U.S. Department of Commerce.

Bajpai, A.C., I.M. Calus, and J.A. Fairley (1977) *Numerical Methods for Engineers and Scientists*, John Wiley, New York.

Bergmann, P.G. (1976) *Introduction to the Theory of Relativity*, Dover, New York.

Corbon, H.C. and P. Stehle (1960) *Classical Mechanics*, John Wiley, New York.

Grivet, P. (1972) *Electron Optics*, 2nd edn., Pergamon, Oxford (U.K.).

Hockney, R.W. and J.W. Eastwood (1981) *Computer Simulation Using Particles*, McGraw-Hill, New York.

Jackson, J.D. (1975) *Classical Electrodynamics* 2nd edn., J. Wiley, New York.

Kirstein P.T., G.S. Kino, and W.E. Waters (1967) *Space Charge Flow*, McGraw-Hill, New York.

CHAPTER 4

METHODS OF CALCULATING ELECTROMAGNETIC FIELDS

We have seen in the preceding chapter that the equations of motion (trajectory equations) can be solved in principle if the guiding electromagnetic fields are known. In this chapter we discuss the calculation of these fields from given electromagnetic sources in specified geometries. We observe the expected fact that very few situations allow for an exact analytical solution of the equations. This section is organized into a succession of analytical and aproximate methods that increase in complexity. The starting point for electromagnetic static situations are the following Maxwell equations (Stratton 1941):

$$\begin{aligned} \vec{\nabla} \times \vec{E} &= 0 , \qquad & \vec{\nabla} \cdot \vec{B} &= 0 , \\ \vec{\nabla} \cdot \vec{D} &= \omega , \qquad & \vec{\nabla} \times \vec{H} &= \vec{J} . \end{aligned} \tag{4.1}$$

In order to avoid confusion with the cylindrical radial coordinate ρ, we will designate the *volume charge density* (in C/m^3) by the symbol ω. The fourth equation above contains the *volume current density* $\vec{J}$. The two source distributions give rise to four electromagnetic fields $\vec{E}$, $\vec{D}$, $\vec{B}$, and $\vec{H}$ that are connected two by two means of the *constitutive relationships*. These are generally quite complicated (Landau and Lipschitz 1960), but for linear isotropic media they are,

$$\vec{D} = \varepsilon \vec{E} , \qquad \vec{H} = \mu^{-1} B . \tag{4.2}$$

It suffices in this work to characterize dielectric materials by their permittivity $\epsilon = \epsilon_r \epsilon_0$ ($\epsilon_0 \approx 8.854 \times 10^{-12}$ F/m), and magnetic materials by their *permeability* $\mu = \mu_r \mu_0$ ($\mu_0 = 4\pi \times 10^{-7}$ H/m). The relative quantities ϵ_r and μ_r are dimensionless. The most important fact about the static Maxwell equations (4.1) is the complete decoupling between electrostatic and magnetostatic fields; ω gives rise to purely electric fields, and $\vec{J}$ to purely magnetic fields. In principle, it is conceivable that a beam of electrons guided by electrostatic electron lenses (i.e., by metallic structures consisting of pieces held at different but fixed voltages) will produce a $\vec{B}$ field which, by Ampère's law, will create currents in the electrodes with magnetic fields working back on the electron beam, but such effects are almost always negligible.

4.1 ELECTROSTATIC FIELDS

In the following sections we first discuss methods of solving for electrostatic fields, and then for magnetic fields, with most stress upon the electrostatic problems.

4.1.1 The Simplest Analytical Solutions

Analytical solutions for $\vec{E}$ fields exist only for a number of "canonical" problems, most of which can be found in any of the plethora of undergraduate texts on electromagnetics, e.g., Cheng (1989), Hayt (1989), and Kraus (1984). In the case of uniform permittivity ϵ in some region of interest, the electrostatic equations of (4.1) are easily combined by defining an electrostatic potential Φ to form *Poisson's equation*,

$$\Delta\Phi = -\frac{\omega}{\epsilon}, \tag{4.3a}$$

which reduces to *Laplace's equation* in source-free regions,

$$\Delta\Phi = 0, \tag{4.3b}$$

given that $\vec{E} = -\vec{\nabla}\Phi$. In electron optics, this latter partial differential equation has solutions determined by values of the potential Φ on some boundaries and by values of the normal components $\hat{n} \cdot \vec{E}$ on other boundaries ($\hat{n}$ is the unit vector normal to the surface at the point in question). Solutions are easily obtained for parallel-plate, coaxial-cylinder, and concentric-sphere capacitors. Layers with differing ϵ can be included if these have sufficient symmetry (e.g., cylindrical layers in cylindrical geometry). Specific solutions for these simplest cases are

$$\vec{E} = \left(\frac{V}{d}\right)\hat{z}\,, \qquad \text{parallel-plate (Fig. 4-1)} \tag{4.4}$$

$$\vec{E} = \left[\frac{V}{\rho}\ln\left(\frac{b}{a}\right)\right]\hat{\rho}\,, \qquad \text{coaxial cylinders (Fig. 4-2)} \tag{4.5}$$

$$\vec{E} = \left[\frac{abV}{(b-a)r^2}\right]\hat{r}\,, \qquad \text{concentric spheres (Fig. 4-3)} \tag{4.6}$$

In all three cases, a voltage V is maintained between the two perfect conductors defining these capacitors, and inclusion of layers (as men-

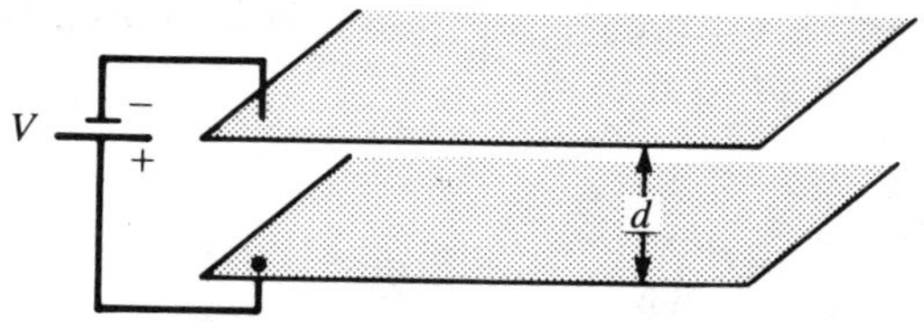

Fig. 4-1. A parallel-plane capacitor.

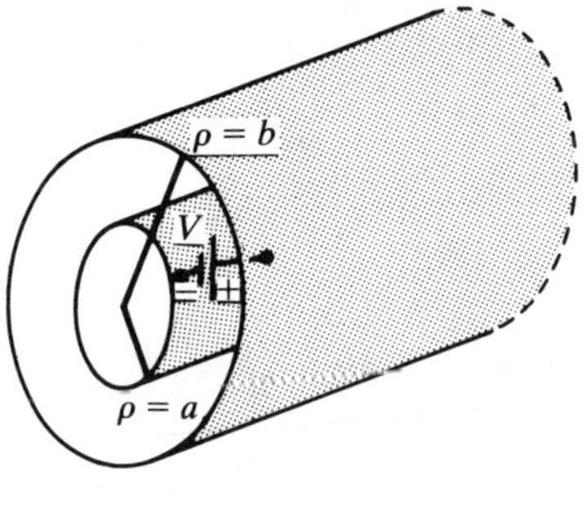

Fig. 4-2. A co-axial circular-cylindrical capacitor.

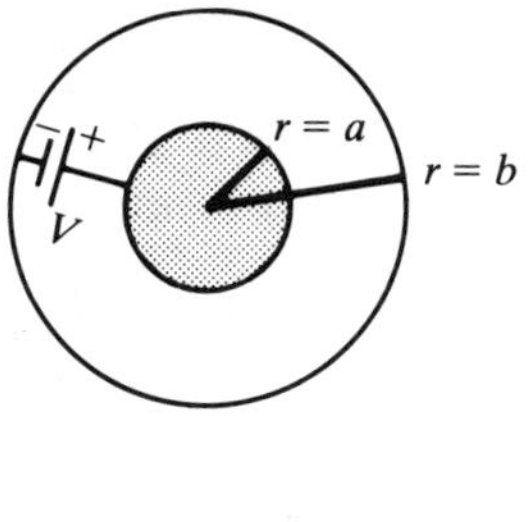

Fig. 4-3. A concentric spherical capacitor.

tioned above) leads merely to fairly straightforward extensions of these formulas for $\vec{E}$ adapted to boundary conditions between layers. There are more complicated geometries that lend themselves to exact analytical solutions of (4.3). Some of these are for specific classes of electron lenses (aperture lenses, cylinder lenses without air gaps, etc.) and these are discussed in a later chapter. Others can be extracted by special techniques (see later sections).

4.1.2 Separation-of-Variable Techniques

In some geometries, a series solution can be extracted from Laplace's equation for the electrostatic potential Φ by means of exploiting a separation of the equation into a sum of component equations; each in less than three variables (most commonly, each in one variable only). The technique is illustrated here for Cartesian coordinates, i.e., for those geometries where potential or normal field boundaries are $x = \text{constant}$, $y = \text{constant}$, and/or $z = \text{constant}$ planes. The solution is then attempted by postulating that $\Phi(x, y, z) = X(x)Y(y)Z(z)$, so that Eq. (4.3) can be rewritten as

$$\frac{d^2X}{X\,dx^2} + \frac{d^2Y}{Y\,dy^2} + \frac{d^2Z}{Z\,dz^2} = 0\,, \tag{4.7}$$

provided $X(x)$, $Y(y)$, and $Z(z)$ are nonzero at locations $\vec{r} = (x, y, z)$ where Eq. (4.7) is presumed to be valid. The only way that (4.7) can be satisfied is by simultaneous solution of the three second-order total differential equations,

$$\frac{d^2X}{dx^2} + k_1^2X = 0\,, \qquad \frac{d^2Y}{dy^2} + k_2^2Y = 0\,, \qquad \frac{d^2Z}{dz^2} + k_3^2Z = 0\,, \tag{4.8}$$

subject to the constraint that the three complex constants k_1, k_2, and k_3 obey

$$k_1^2 + k_2^2 + k_3^2 = 0\,. \tag{4.9}$$

Two of the three k_i $(i = 1, 2, 3)$ are usually determined by boundary conditions, and the third is then determined by Eq. (4.9). Each of the equations in (4.8) can be written as a similar one in $X_i(x_i)$, and the general solution is

$$X_i(x_i) = A_i \sin(k_ix_i) + B_i \cos(k_ix_i)\,, \tag{4.10}$$

where A_i and B_i are two complex constants. The function $X_i(x_i)$ is a combination of trigonometric functions if k_i is real, and it is a combination of hyperbolic (or exponential) functions for imaginary k_i. It is linear in x_i when $k_i = 0$. It is usually a mixture of trigonometric and hyperbolic functions for complex k_i, but k_i is either real or imaginary in most applications where separation of variables is possible. In general, the solution for $X_i(x_i)$ will involve many wavenumbers $k_{i1}, k_{i2}, k_{i3}, \ldots$. To see why, consider the example in Fig. 4-4 of a rectangular parallelopipedum (e.g., a rectangular waveguide) of dimensions $0 < x < a$ and $0 < y < b$ with $\Phi(x, y)$ being some function of x and y at $z = 0$, and with $\Phi = 0$ on the four side plates at $x = 0$, $x = a$, $y = 0$, $y = b$ (we leave the limit $z \to \pm\infty$ undefined). Consideration of the symmetries will lead one to choose $X(x) \propto \sin(m\pi x/a)$ and $Y(y) \propto \sin(n\pi y/b)$, so that we have chosen $k_1 = m\pi/a$ and $k_2 = n\pi/b$, where m, n are positive integers. The constraint (4.9) then determines k_3 to be

$$k_3 = - \pm i\pi\sqrt{\frac{m^2}{a^2} + \frac{n^2}{b^2}}\,, \tag{4.11}$$

and the reasonable requirement that $\Phi(x, y, z) \to 0$ as $z \to \infty$ tells us that $Z(z) \propto \exp(k_3 z)$ for $z > 0$. The boundary condition at $z = 0$ is then

$$\Phi(x, y, 0) = V_{mn} \sin\left(\frac{m\pi x}{a}\right) \sin\left(\frac{n\pi y}{b}\right), \tag{4.12}$$

where V_{mn} is a constant. This condition cannot be fulfilled, unless $\Phi(x, y, 0)$ was already chosen, fortuitously, to be proportional to the product of the two sines in Eq. (4.12). However, regardless of whatever boundary condition $\Phi(x, y, 0)$ needs to be met, we can always attempt to fit the superposition,

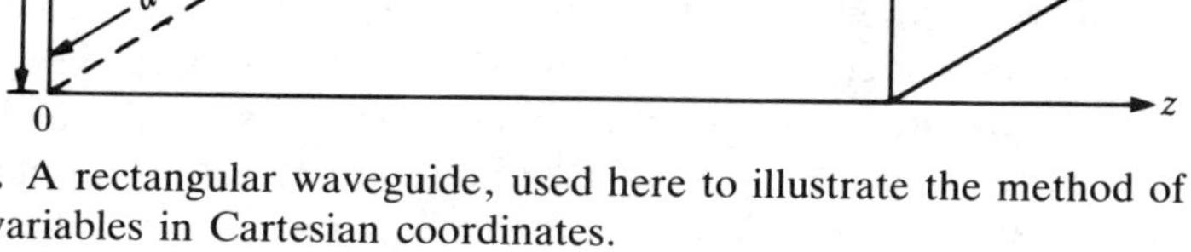

Fig. 4-4. A rectangular waveguide, used here to illustrate the method of separation-of-variables in Cartesian coordinates.

$$\Phi(x, y, z) = \sum_{m=1}^{\infty} \sum_{n=1}^{\infty} V_{mn} \sin\left(\frac{m\pi x}{a}\right) \sin\left(\frac{n\pi y}{b}\right) \exp(k_{mn} z) ,$$
$$k_{mn} = -\sqrt{k_{1m}^2 + k_{2n}^2} = -\sqrt{\frac{m^2}{a^2} + \frac{n^2}{b^2}} , \tag{4.13}$$

with $k_1 = k_{1m} = m\pi/a$, $k_2 = k_{2n} = n\pi/b$, and $k_3^2 = -k_1^2 - k_2^2$ for each term. The boundary condition at $z = 0$ must now be fitted to

$$\Phi(x, y, 0) = \sum_{m=1}^{\infty} \sum_{n=1}^{\infty} V_{mn} \sin\left(\frac{m\pi x}{a}\right) \sin\left(\frac{n\pi y}{b}\right) , \tag{4.14}$$

which yields a unique solution for all V_{mn} by means of the theory of Fourier series, given that $\Phi(x, y, 0)$ is reasonably well behaved (e.g., is continuous and differentiable). The Fourier inversion of Eq. (4.14) then yields

$$V_{mn} = \frac{4}{ab} \int_0^a dx \int_0^b dy \, \Phi(x, y, 0) \sin\left(\frac{m\pi x}{a}\right) \sin\left(\frac{n\pi y}{b}\right) . \tag{4.15}$$

Equation (4.13) represents a series solution for the problem illustrated in Fig. 4-4, and it is only one of a number of problems that is amenable to this type of solution.

The above example illustrates the general procedure of separation-of-variables methods. We have noted from the example that symmetry of the surface boundaries of the volume in which Laplace's equation is to be solved is required with respect to the coordinate system to be used. Procedures similar to the one above can be followed for cylindrical, for spherical, and for a number of other coordinate systems that allow a separation of the Laplacian operator (Stratton 1941). A generalization of the method of separation of variables can lead to two-dimensional Fourier integrals instead of series as in Eq. (4.13). Some examples are given by de Wolf (1978, 1981).

4.1.3 Method of Images

A class of solutions to Laplace's equation (4.3b) may be obtained by the *method of images* when the situation of charge sources and boundary surfaces possesses sufficient symmetries. The method replaces a system of charges and bounding surfaces with a much simpler system which yields the same potentials inside a specified volume of interest. The method is based on the fact that solutions to Laplace's equation are unique when

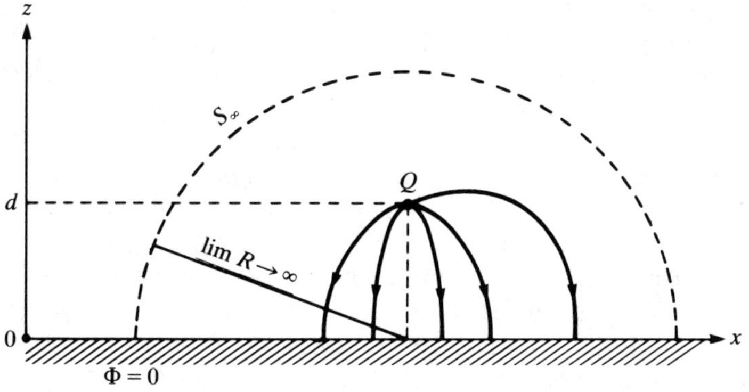

(4.5a)

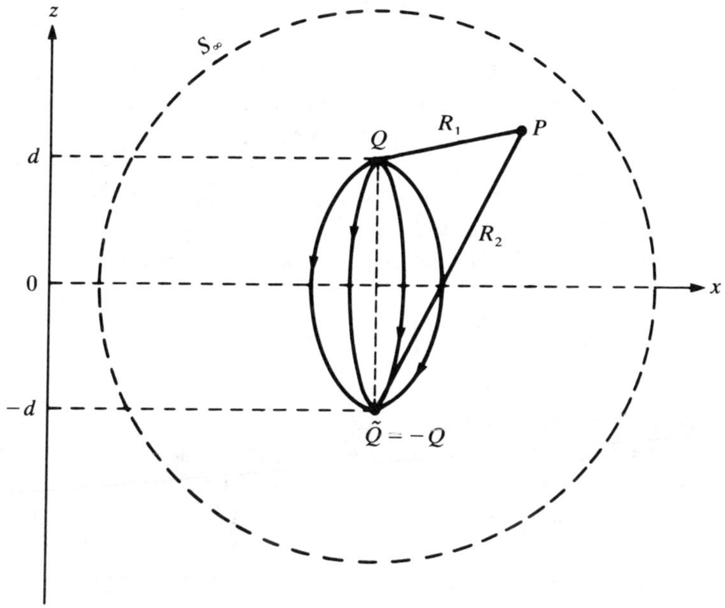

(4.5b)

Fig. 4-5. (a) A line charge above a grounded perfectly-conducting plane (at $\Phi = 0$ volts). Several field lines are shown. (b) The situation of (a) but with an image point $\tilde{Q}$ replacing the grounded plane.

the volume in question is bounded by surfaces on which either potential values $\Phi(\vec{r}_s)$ or normal derivatives $\hat{n} \cdot \vec{E}(\vec{r}_s)$ are specified. In the second case, the solution is unique in Φ except for a constant. A third case yielding unique solutions is a combination: $\Phi(\vec{r}_s)$ is specified on parts of the bounding surface and $\hat{n} \cdot \vec{E}(\vec{r}_s)$ on the remaining parts. The bounding surface may well include S_∞ (the surface at infinity), i.e., it may be that potential or field is specified at infinity over part of the region.

A classical example is sketched in Fig. 4-5a. A charge Q is at height $z = d$ above a perfectly conducting plane at $z = 0$. We have sketched the hemispherical infinite surface S_∞ above this plane as a half-spherical surface at a finite distance R, even though we intend to convey that we will let $R \to \infty$ when dealing with it. Electric field lines are perpendicular to the plane $z = 0$ as it is an equipotential. We expect that $\Phi(\vec{r}) \propto 1/R$ as the distance R of location $\vec{r}$ increases from the location of the charge, and thus that $\Phi(\vec{r}_s)$ on S_∞ will be zero. The electric field lines are as sketched, and an analytical expression seems intractable. The method of images suggests that the charge and zero-potential plane of Fig. 4-5a be replaced by the two charges of Fig. 4-5b. This latter situation consists of one charge Q at $z = d$ and an image charge $\tilde{Q} = -Q$ at $z = -d$. In *both* figures there is a charge Q at $z = d$ and the surface $z = 0$ is an equipotential with $\Phi = 0$ (as is easily seen from Eq. 4.16 for the image situation). Moreover, $\Phi(\vec{r}_s) = 0$ on the upper hemisphere S_∞ for both situations. Hence the solutions to *both* problems in the region $z > 0$ must be identical. The advantage (and beauty!) of the two-charge situation is that we can express the potential $\Phi(\vec{r})$ in the upper half-plane easily by virtue of the superposition of potentials of two point charges as

$$\Phi(\vec{r}) = \frac{Q}{4\pi\epsilon_0}\left(\frac{1}{R_1} - \frac{1}{R_2}\right). \tag{4.16}$$

The method of images does not allow us to say anything about the situation at $z < 0$ for Fig. 4-5a; there is no relationship whatever between Eq. (4.16) and the lower hemisphere of Fig. 4-5a. This typical application of the method of images is intended only for the upper hemisphere.

A number of other relatively simple situations are amenable to similar analyses by means of image charges. Examples include a charge inside a two-plane wedge with an angle between the two planes such that 360° is an integer multiple of it (and with these planes held at one equipotential, also a charge sandwiched between two infinite parallel grounded planes, a point charge and an infinite circular cylinder or sphere held at an equipotential, and perhaps several other similar situations. Most of these can be found in any undergraduate text on electromagnetics; an older reference is Smythe (1969).

4.1.4 Schwarz-Christoffel Transformation

Whenever a region has fields and potentials which are two-dimensional in planar geometry, e.g., have no dependence upon the Cartesian y coordinate, and whenever the bounding surfaces of such a region are characterized by a number of planes $a_1x + a_2z = a_3$ so that the intersections with $y = 0$ form a polygon (some sides of which might very well include S_∞) (see Fig. 4-6) then a *Schwarz-Christoffel transformation* may enable one to obtain an analytical solution for the equipotential planes (or lines in two dimensions) inside the polygon.

The Schwarz-Christoffel transformation replaces the polygon by a much simpler one through a transformation; the much simpler polygon allows one to write down the equipotential planes (lines in the $y = 0$ plane) immediately. The remaining problem, not always trivial, lies in transforming back to the original situation. If it is successful, then a valid solution of Laplace's equation has been obtained! One might think of this procedure as being analogous to Fourier transformation of a differential equation in one independent variable; the Fourier-transformed equation becomes purely algebraic, and the remaining problem, also not always trivial, is to perform the inverse Fourier transformation upon the algebraic solution.

The Schwarz-Christoffel transformation is essentially an application of the more general class of *conformal transformations* of a function of a complex variable. Let us first review conformal transformations very briefly. Define the complex variable $\zeta \equiv x + iz$ (where $i \equiv \sqrt{-1}$). Let $w = f(\zeta)$ represent a mapping of points of the complex-ζ plane into points of a complex-w plane; $f(\zeta)$ is a *meromorphic* function (a function that is analytic except at certain points). In the analytic region of $f(\zeta)$ we have

$$dw = f'(\zeta)\, d\zeta\ , \tag{4.17}$$

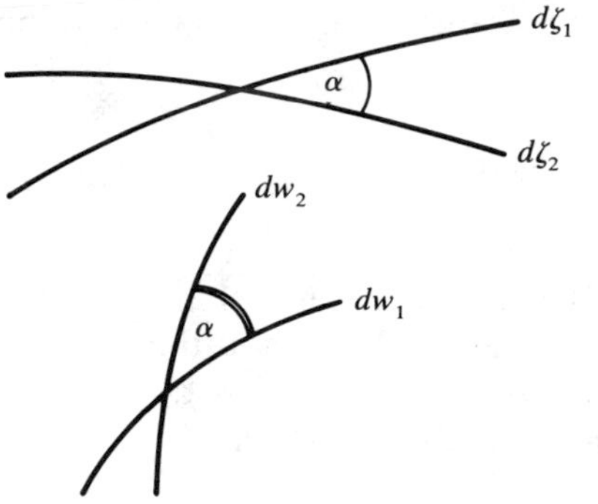

Fig. 4-6. Conservation of angle between lines in the ζ and in the w planes for conformal transformations $w = f(\zeta)$ in regions where $f(\zeta)$ is analytic in ζ.

where $f'(\zeta)$ is independent of the direction of $d\zeta$ in the ζ plane. Write $f'(\zeta)$ in polar form as

$$f'(\zeta) = R(\zeta)\exp[i\psi(\zeta)]\,. \tag{4.18}$$

Both $R(\zeta)$ and $\psi(\zeta)$ are independent of the direction of $d\zeta$ in the ζ plane. Hence if two infinitesimal line segments in the ζ plane, as depicted in Fig. 4-6, namely $d\zeta_1$ and $d\zeta_2$ intersect at an angle α, then we can write

$$d\zeta_2 = d\zeta_1 \exp(i\alpha)\,. \tag{4.19}$$

From Eqs. (4.17) and (4.18) it follows that the corresponding relationships for dw is

$$dw_2 = R(\zeta)\exp[i\psi(\zeta)]\,d\zeta_2 = R(\zeta)\exp[i\psi(\zeta)]\,d\zeta_1 e^{i\alpha} = dw_1 \exp(i\alpha)\,. \tag{4.20}$$

The transformed line segments intersect at the same angle α and the transformation or mapping $\zeta \rightarrow w$ is known as a *conformal* one.

The Schwarz-Christoffel transformation attempts to transform a polygon in the $\zeta = x + iz$ plane (as shown in Fig. 4-7a) into a simpler shape in the $w = w_r + iw_i$ plane that can serve as the boundary of a more elementary area (volume in three dimensions) inside which $\Delta\Phi = 0$ is more easily solved than in the original x, z coordinates. The Laplace equation $\Delta\Phi = 0$ remains unchanged in two dimensions when we transform from $\zeta = x + iz$ to $w = w_r + iw_i$. Although that is a relatively elementary result of complex function theory for a function $\Phi(x, z)$ that is analytic in $\zeta = x + iz$ in a specified two-dimensional area, we review the essential steps in proving that in Appendix I.

The polygon in Fig. 4-7a represents a *meromorphic* function $f(\zeta)$ that is not analytic only at the vertices $\zeta_1, \zeta_2, \ldots, \zeta_n$. One particular function that has the right behavior at $\zeta = \zeta_i$ has the transformation property in the immediate vicinity of the vertex,

$$\zeta - \zeta_i = A_i(w - w_i)^{1+p_i}\,, \tag{4.21a}$$

with constant A_i and $-1 < p_i < 0$. This behavior ensures that $d\zeta \propto (w - w_i)^{p_i}\,dw$ in the immediate vicinity of the vertex ζ_i, w_i, and the function clearly cannot be analytic there. An obvious extension of this idea to include all (but one of) the vertices suggests the transform,

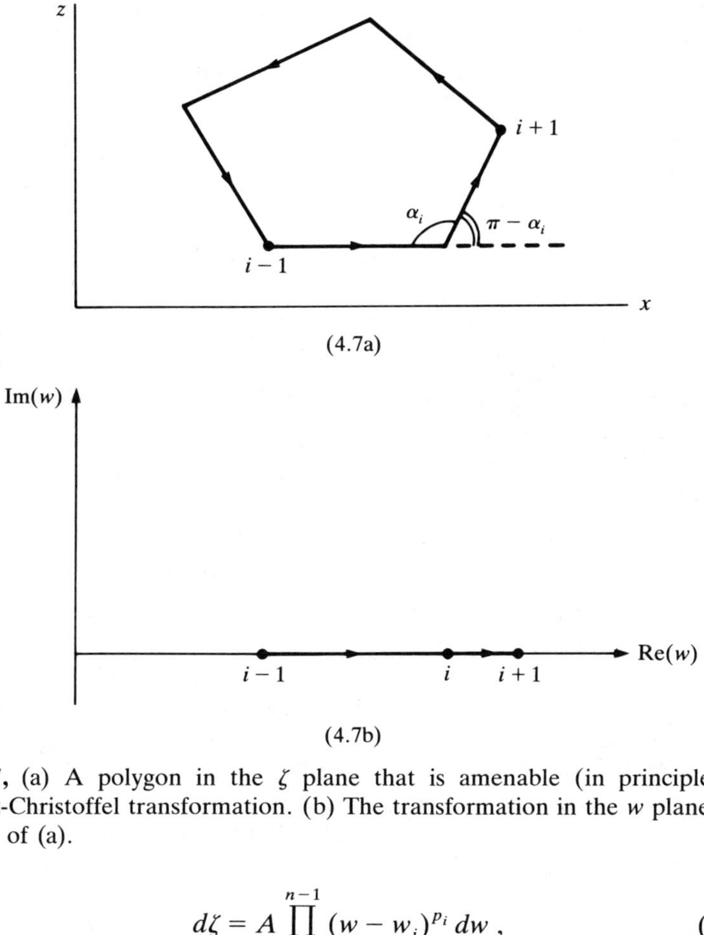

Fig. 4-7, (a) A polygon in the ζ plane that is amenable (in principle) to a Schwarz-Christoffel transformation. (b) The transformation in the w plane of the polygon of (a).

$$d\zeta = A \prod_{j=1}^{n-1} (w - w_j)^{p_i}\, dw \,, \tag{4.21b}$$

with constant A. This transform also represents a function that is analytic everywhere, except at the vertices $\zeta_1, \ldots, \zeta_{n-1}$ of the polygon. We use the property (4.20) to determine the powers p_i, and we explain why the $j = n$ term is missing in Eq. (4.21b).

In the immediate vicinity of the ith vertex we have

$$\begin{aligned} d\zeta &= \lim_{w \to w_i} \left\{ A \prod_{j=1}^{n-1} (w_i - w_j)(w - w_i)^{p_i}\, dw \right\} \\ &= \lim_{w \to w_i} \left\{ B(w - w_i)^{p_i}\, dw \right\} = \lim_{w \to w_i} \left\{ \frac{1}{p_i}\, B(w - w_i)^{1+p_i} \right\} . \end{aligned} \tag{4.22}$$

The application of a limiting procedure here must be viewed with some care, as the function is non-analytic at the limit point $w = w_i$ so that the result depends upon the *direction* in which w approaches w_i. For w lying between w_i and w_{i+1}, Eq. (4.22) will predict

$$d\zeta_2 = \frac{1}{p_i} B_i (dw_2)^{1+p_i} ,$$

whereas for w between w_{i-1} and w_i, it will lead to

$$d\zeta_1 = \frac{1}{p_i} B_i (dw_1)^{1+p_i} .$$

The property (4.20) tells us that $d\zeta_2 = \exp[i(\pi - \alpha_i)]\, d\zeta_1$, and this leads, upon application to the two preceding expressions above for $d\zeta_2$ and $d\zeta_1$, to

$$dw_2 = \exp\left[\frac{i(\pi - \alpha_i)}{1 + p_i}\right] dw_1 . \tag{4.23a}$$

The greatest simplification we can attain is to make $dw_2 = \exp(i\pi)\, dw_1 = -dw_1$, i.e., to choose

$$p_i = -\frac{\alpha_i}{\pi} \quad \text{for } 1 \leqslant i \leqslant n - 1 . \tag{4.23b}$$

As a result, the bounding sides of the polygon in the ζ plane are mapped into a straight line in the w plane, for which we may choose the $\mathrm{Re}(w) = w_r$ axis, as shown in Fig. 4-7b. Subsequent integration of Eq. (4.21b) with the values (4.23b) yields

$$\zeta = \zeta_0 + AF(w_1, \ldots, w_{n-1}) + C , \tag{4.24}$$

where $F(w_1, \ldots, w_{n-1})$ is a complex function resulting from indefinite integration of (4.21b), C is a real constant, A and ζ_0 are complex constants that result from indefinite integration. As there are n vertices, we see that the left-hand side of (4.24) has $2n$ real constants. However, there are $2n + 3$ real constants on the right-hand side! We still have the liberty of choosing three vertices arbitrarily on the Re(w) axis. It is customary to choose w_n at $w \rightarrow \pm\infty$, which is why it need not appear explicitly in (4.21b). The other two w_i vertices may be chosen as conveniently as possible for the problem at hand. We observe that motion from $\zeta_{i-1} \rightarrow \zeta_i \rightarrow \zeta_{i+1}$ corresponds to a motion on the Re(w) axis towards the right: $w_{i-1} \rightarrow w_i \rightarrow w_{i+1}$, as shown in Fig. 4-7b explicitly by arrows.

The result of the transformation is to map the polygon sides into portions of the Re(w) axis. When Dirichlet boundary conditions (values of Φ on the bounding surfaces) are specified, the ensuing problem in the w plane is often easier to solve. If, for example, $n = 2$ (a wedge), then w_1 can be placed at the origin and w_2 is at Re(w) $= \pm\infty$. With V_2 from $-\infty$ to 0, and V_1 from 0 to ∞ on the Re(w) axis, the solution for potential $\Phi(w)$ in the upper half-plane is a fan of straight-line equipotentials: $\Phi(\phi) = V_1 + (V_2 - V_1)(\phi/\pi)$, as illustrated in Fig. 4-8. The inverse transformation of $\Phi(\phi)$ to the ζ plane can be found from Eq. (4.21) by standard contour methods of complex-function theory.

As with all of the preceding methods, conformal transformations will enable one to obtain potentials $\Phi(\vec{r})$ only for a relatively restricted class of problems. Significant generality seems to be possible only through *numerical* solution of Laplace's or Poisson's equation.

4.1.5 Numerical Methods

In almost all electron optics problems, the right-hand side of Poisson's equation is the volume charge density of electrons in the beam that is otherwise guided by electromagnetic fields created in turn by metallic grids and plates held at fixed voltages, or by wires with mixed currents. The electromagnetic field is therefore a superposition of solutions of Laplace's equation (4.3b) and of fields produced by the beam itself. We shall see that the *space-charge* fields due to the beam can be handled

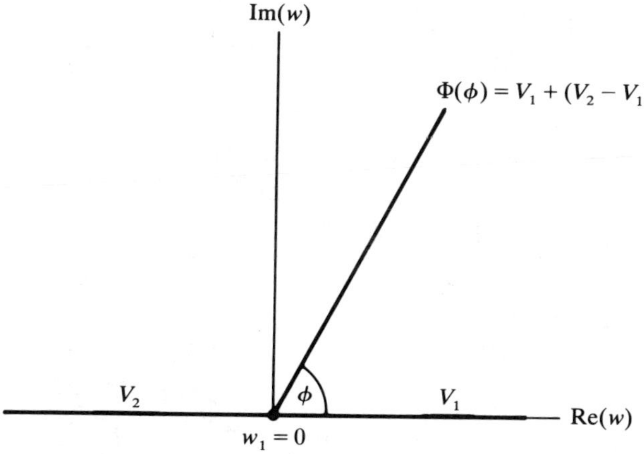

Fig. 4-8. The solution in the w plane for equipotentials in a mapping that yields equipotential values V_1 on the Re(w) > 0 axis and V_2 on the Re(w) < 0 axis.

separately; here, we observe that the remaining fields are solutions of Laplace's equation.

This section is concerned with numerical solutions of Eq. (4.3b). There are several numerical methods that underlie most computer-generated simulations of electrostatic fields: *finite-mesh relaxation* methods and *integral-equation* methods. Each has advantages and disadvantages, and use is dictated by the available boundary conditions.

4.1.5.1 Finite-Mesh Relaxation Methods. The basic idea is to discretize a volume bounded entirely by a surface on which $\Phi(\vec{r}_s)$ is known (Dirichlet boundary conditions) into a three-dimensional grid of mesh points. If the grid spacing (or mesh-size parameter) h is sufficiently small then Laplace's equation can be approximated to good accuracy by a difference equation.

We will illustrate the method for a rectangular parallelopipedum with voltages $V_1, V_2, \ldots, V_6$ on the six sides. The solution to $\Phi(\vec{r})$ in the interior is obtained from a superposition of six problems, each with the same parallelopipedum geometry, but with only one nonzero voltage V_i $(1 \leqslant i \leqslant 6)$ on a side. Consider the interior point $\vec{r} = (x, y, z)$ and its six nearest neighbors as sketched in Fig. 4-9. We characterize the nearest-neighbor potentials as $\Phi(x, y, z + h) = \Phi_1$, $\Phi(x, y, z - h) = \Phi_2$, and four other obvious designations. Let the central point have potential $\Phi(\vec{r}) \equiv \Phi_0$ for short. A Taylor-series expansion yields for Φ_1 and Φ_2,

$$\Phi_1 = \Phi_0 + h\,\frac{\partial \Phi_0}{\partial z} + \frac{1}{2}\,h^2\,\frac{\partial^2 \Phi_0}{\partial z^2} + \frac{1}{6}\,h^3\,\frac{\partial^3 \Phi_0}{\partial z^3} + \frac{1}{24}\,h^4\,\frac{\partial^4 \Phi_0}{dz^4} + \cdots,$$

$$\Phi_2 = \Phi_0 - h\,\frac{\partial \Phi_0}{\partial z} + \frac{1}{2}\,h^2\,\frac{\partial^2 \Phi_0}{\partial z^2} - \frac{1}{6}\,h^3\,\frac{\partial^3 \Phi_0}{\partial z^3} + \frac{1}{24}\,h^4\,\frac{\partial^4 \Phi_0}{dz^4} + \cdots. \tag{4.24}$$

The sum of these two nearest neighbors yields

$$\Phi_1 + \Phi_2 = 2\Phi_0 + h^2\,\frac{\partial^2 \Phi_0}{\partial z^2} + \frac{1}{12} h^4\,\frac{\partial^4 \Phi_0}{\partial z^4} + \cdots. \tag{4.25}$$

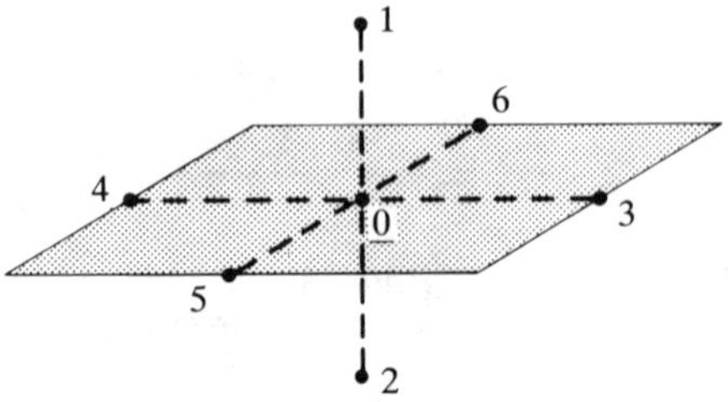

Fig. 4-9. A rectangular discretization showing the six nearest neighbors of a mesh point.

Similar Taylor expansions can be made for the four other nearest neighbors (two by two), and if these are all added to (4.25) there results a sum of the six nearest neighbors on the left-hand side, and Φ_0 plus derivatives on the right-hand side:

$$\sum_{i=1}^{6} \Phi_i = 6\Phi_0 + h^2 \, \Delta\Phi_0 + \frac{h^4}{12}\left(\frac{\partial^4\Phi_0}{\partial x^4} + \frac{\partial^4\Phi_0}{\partial y_4} + \frac{\partial^4\Phi_0}{\partial z^4}\right) + \cdots . \tag{4.26}$$

Because Laplace's equations predicts that $\Delta\Phi_0 = 0$, it follows that

$$\Phi_0 = \frac{1}{6}\sum_{i=1}^{6} \Phi_i\left[1 + \mathrm{O}\left(\frac{h^4}{l^4}\right)\right], \tag{4.27}$$

where l^4 is the fourth power of a length defined by the fourth derivatives in Eq. (4.26). This error needs to be modified in practice (see below). Thus, each mesh point has a potential $\Phi_0(x, y, z)$ defined by the arithmetic average of its nearest neighbors to $\mathrm{O}(h^4/l^4)$ accuracy. In a two-dimensional geometry, the factor 1/6 in Eq. (4.27) becomes 1/4. Consider for simplicity of representation the two-dimensional sketch of Fig. 4-10. There are six interior points for which Φ_i (for $1 \leqslant i \leqslant 6$) must be determined. If V_1, V_2, V_3, and V_4 are given as shown, then the equations are

$$\begin{aligned}
4\Phi_1 &= V_3 + \Phi_4 + V_4 + \Phi_2 ,\\
4\Phi_2 &= V_3 + \Phi_5 + \Phi_1 + \Phi_3 ,\\
4\Phi_3 &= V_3 + \Phi_6 + V_2 + \Phi_2 ,\\
4\Phi_4 &= V_1 + \Phi_1 + V_4 + \Phi_5 ,\\
4\Phi_5 &= V_1 + \Phi_2 + \Phi_4 + \Phi_6 ,\\
4\Phi_6 &= V_1 + \Phi_3 + V_2 + \Phi_5 ,
\end{aligned} \tag{4.28}$$

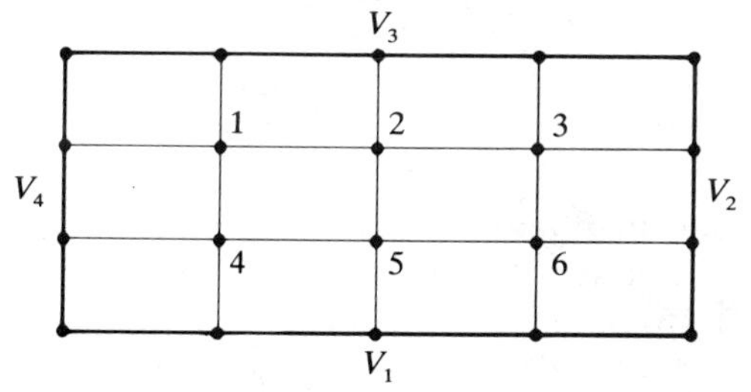

Fig. 4-10. A two-dimensional rectangular Dirichlet boundary-value problem for which the interior potentials are easily obtained by a finite-relaxation method of Laplace's equation.

if the error terms $\propto O(h^4/l^4)$ are ignored. These are six linear inhomogeneous equations in six unknowns which have a unique solution (as the determinant of the properly ordered matrix is nonzero). In three dimensions and for N interior points, a similar procedure will result in N linear inhomogeneous equations (each equation containing at most seven unknowns).

If the right-hand side of Eq. (4.27) contains errors, e.g., as would occur using an iterative method (see below) to determine the simultaneous solution of the N linear equations in the interior potentials, then it can be shown (Weber 1967) that the error is $O(h^2)$ rather than $O(h^4)$. Truncation and roundoff errors occur (aside from computer roundoff errors) because each equation (4.27) for an interior point is expressed as a linear sum of other interior (plus perhaps some exterior) points, whereas $O(h^4)$ holds for Φ_0 only if the nearest-neighbor Φ_i are already all exact solutions of a similar equation. The accumulation of such errors increases $O(h^4)$ to $O(h^2)$. Equations similar to (4.27) but with other numerical coefficients for the Φ_i can be found for cylindrical, spherical, or other geometrical configurations (Weber 1967).

The solution of N linear equations in N unknowns with a nonzero inhomogeneous term is theoretically trivial, but requires close attention in practice because the major step is the inversion of an $N \times N$ matrix. Limitations in computer storage ability may render the usual methods (Bonjour 1980) invalid, even though this matrix is sparse when N is large (for a three-dimensional grid there are no more than seven nonzero elements on each row of the matrix).

A *relaxation* method overcomes the storage problem easily; it requires only $2N$ storage locations (perhaps $3N$ with error criterion), rather than N^2 in the computer memory. In relaxation, an iterative procedure is followed. An initial, even arbitrary, guess is made for the N interior potential values $\Phi_i^{(0)}$. A series of higher-order estimates of the N interior potential values are made by means of the recursion relationship,

$$\Phi_{i0}^{(n+1)} = \frac{1}{6} \sum_{k=1}^{6} \Phi_{ik}^{(n)} , \tag{4.29}$$

which is merely an adaptation of Eq. (4.27) to Φ_{i0} and its six nearest neighbors Φ_{ik} (these latter chosen in iteration $\#n+1$ from the results of iteration $\#n$). The process can be shown to be convergent under almost universal circumstances (Weber 1967), and it can be shortened by adjusting each newly obtained Φ_i by a process of *over-relaxation*, i.e., by replacing $\Phi_i^{(n+1)}$ by $\tilde{\Phi}_i^{(n+1)}$ via

$$\tilde{\Phi}_i^{(n+1)} = \Phi_i^{(n)} + \omega[\Phi_i^{(n+1)} - \Phi_i^{(n)}] \,. \tag{4.30}$$

Initial choice of over-relaxation parameter ω larger than 1 will move the iterated value more rapidly towards its asymptotic value as $n \rightarrow \infty$. If ω is chosen larger than 1 relatively further on in the iteration procedure, it may delay mattters as there will then be successive over- and undershooting of the asymptotic value. There are systematic methods of deciding an optimal value of ω (Forsyth and Wasov 1960, Varga 1971), but these are difficult in practice, and a reasonable engineering practice is to allow ω to be an adjustable parameter to be chosen by trial and error and/or experience for a particular class of geometries (Bonjour 1980). The iteration procedure remains quite robust; for large enough values of n the solutions will become arbitrarily close to the exact values of the *approximate* equations. We need to stress that the approximate equation (4.27) leads to exact solutions that are not exact solutions of Laplace's equation. The difference lies in error terms of $O(h^2/l^2)$. However, by decreasing the mesh parameter h and redoing the iteration procedure, one can converge upon arbitrarily precise solutions of Laplace's equation. It is obvious that computer time is the price to be paid for greater accuracy from smaller mesh parameters, and the usual engineering practice is to strike a balance between desired accuracy and available resources.

Finite-mesh relaxation methods can also be adapted to Neumann boundary conditions (specification of $\hat{n} \cdot \vec{E} = \omega_s/\epsilon_0$, where ω_s is the surface-charge density in the case of a perfect-conductor boundary), but this is not discussed further.

4.1.5.2 Integral Equation Methods. An entirely different numerical approach is based upon the fact that electrostatic field lines terminate perpendicularly on perfect-conductor surfaces: at the termination we have the boundary condition $\hat{n} \cdot \vec{E} = \omega_s/\epsilon_0$, where $\hat{n}$ is the unit vector normal to the surface, and ω_s is the surface-charge density (measured in C/m^2). Whatever the "man-made" cause of the electrostatic fields in the regions between and outside the bounding conductor plates, we can consider $\omega_s(\vec{r}_s)$ on the surfaces of the conductors (as defined by $\hat{n} \cdot \vec{E}$) to be the sources of these fields (see Fig. 4-11). Referring back to the discussion in Section 4.1.3 on uniqueness of potential solutions in electrostatics, we observe that specification of all $\omega_s = \epsilon_0 \hat{n} \cdot \vec{E}$ suffices to guarantee a unique solution for the interior $\vec{E}$ fields of the region. The boundary may well include parts of S_∞ (where, often but not always, $\omega_s = 0$).

We obtain values of the potential $\Phi(\vec{r})$ in the interior from knowledge of the charge density ω_s on all boundary surfaces, using the integral expression

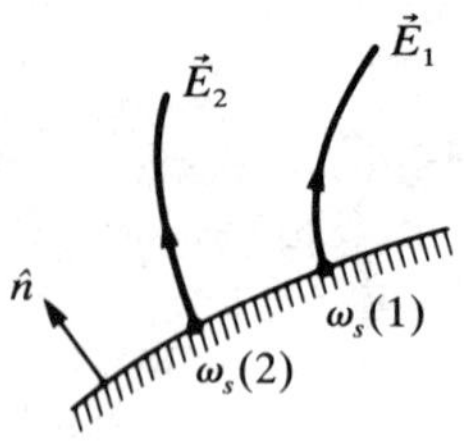

Fig. 4-11. Surface charge on and field lines ending on the surface of a perfect conductor.

$$\Phi(\vec{r}_P) = \frac{1}{4\pi\epsilon_0} \int_S \int dS_Q \, \frac{\omega_s(\vec{r}_Q)}{R_{PQ}} , \tag{4.31}$$

where R_{PQ} is the distance between source point $\vec{r}_Q$ and field point $\vec{r}_P$ (see Fig. 4-12). Equation (4.31) can be discretized by partitioning (see Fig. 4-13) surface S into small surface elements S_Q such that $\omega_s(\vec{r}_Q)$ is approximately constant over the area S_Q of each element and we assign the constant value ω_{sQ} to the entire surface element. Equation (4.31) is then approximated by a discrete sum,

$$\begin{aligned} \Phi(\vec{r}_P) &\approx \sum_{Q=1}^{N} K_Q(\vec{r}_P)\omega_{sQ} , \\ K_Q(\vec{r}_P) &= \frac{1}{4\pi\varepsilon_0} \int_{S_Q} \int dS \, \frac{1}{|\vec{r}_P - \vec{r}_S|} \end{aligned} \tag{4.32}$$

Figure 4-13 depicts the geometry: $\vec{r}_Q$ is a more or less arbitrarily chosen

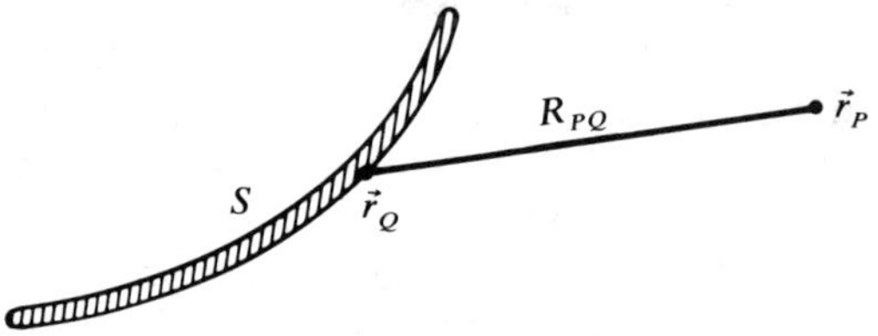

Fig. 4-12. The geometry involved in Eq. (4.31) for the potential at a field point $\vec{r}_P$.

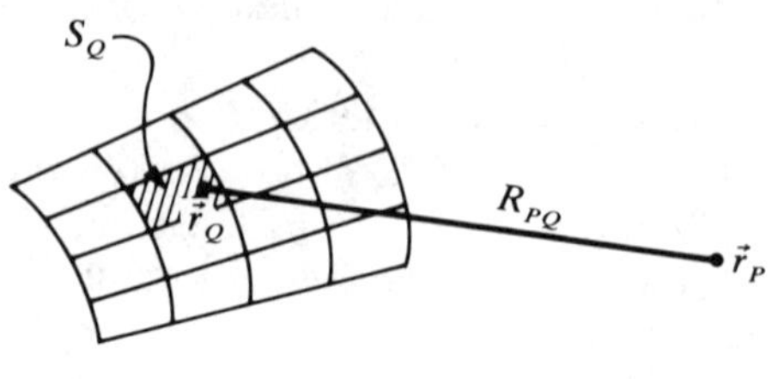

Fig. 4-13. Discretization of the surface-charge distribution in the integral-equation method.

central point of S_Q and R_{PQ} is the distance between it and the field point $\vec{r}_P$. The surface integral in the second of Eqs. (4.32) is over all locations of $\vec{r}_S$ on the surface element with area S_Q. Evaluation of $\Phi(\vec{r}_P)$ from Eq. (4.32) requires the ability to evaluate geometrical expressions and surface-charge densities as follows:

1. We need expressions for $K_Q(\vec{r}_P)$. At very large distances we may well be able to approximate,

$$K_Q(\vec{r}_P) \approx \frac{S_Q}{4\pi\varepsilon_0 R_{PQ}} \; . \tag{4.33}$$

but at intermediate distances we will require more accurate expressions which derive from expansions of $1/|\vec{r}_P - \vec{r}_Q|$ to powers of $1/R_{PQ}$, and from other expansions at very short distances. The principle is straightforward and the details are very specific to the geometry of the surface elements; hence there is no reason to delve further into this here.

2. Evaluation of ω_{sQ} at the center points $\vec{r}_Q$ of each of the N surface elements S_Q is also required. To obtain that, we apply a discretization of Eq. (4.31) similar to that of Eq. (4.32), but now for N field points $\vec{r}_Q$ each of which is a center point of an S_Q surface element. If V_Q is the (known) potential of surface element S_Q, then we obtain

$$\begin{aligned} V_Q &\approx \sum_{Q'=1}^{N} K_{QQ'}\omega_{sQ'} \; , \\ K_{QQ'} &= \frac{1}{4\pi\epsilon_0}\int_{S_{Q'}}\int dS_{Q'}\, \frac{1}{|\vec{r}_Q - \vec{r}_{Q'}|} \; . \end{aligned} \tag{4.34}$$

The $K_{QQ'}$ can be evaluated by methods similar to the evaluations of Eq. (4.33), and therefore we have N linear inhomogeneous equations in the N unknowns, $\omega_{sQ_1}, \ldots, \omega_{sQ_N}$, which can be solved uniquely. There is no singularity problem with the coefficient $K_{QQ'}$ when $Q' = Q$ (as one can see easily by making $\vec{r}_Q$ the center of a local coordinate system in surface element S_Q).

Many variations of this method exist; one of the best known is the *method of moments* (Harrington 1968). In the method of moments, Eq. (4.31) is solved somewhat differently. We attempt to find trial functions $\psi_n(\vec{r}_Q)$ over the bounding surface S such that the continuous charge density $\omega_s(\vec{r}_Q)$ is represented well by the weighted sum of N such trial functions:

$$\omega_s(\vec{r}_Q) \approx \sum_{n=1}^{N} a_n \psi_n(\vec{r}_Q) . \tag{4.35}$$

This develops Eq. (4.31) into the weighted sum

$$\Phi(\vec{r}_P) \approx \sum_{n=1}^{N} a_n I_n(\vec{r}_P) , \qquad I_n(\vec{r}_P) = \int_S \int dS_Q \frac{\psi_n(\vec{r}_Q)}{R_{PQ}} . \tag{4.36}$$

In order to solve for the coefficients a_n, the first of Eqs. (4.36) needs to be applied to N choices of point P on the surface S, i.e., to N centers $\vec{r}_Q$ for which N values V_N of the surface potential are known. The coefficients $I_n(\vec{r}_Q)$ must then be evaluated from the second of Eqs. (4.36) by methods similar to those evoked in evaluating Eq. (4.33) or by other methods dictated perhaps by choice of trial function. At any rate, the accuracy requires a good choice of trial functions and an adequate number N of them. A simple example is given by Neff (1987).

4.1.5.3 Integral Equation Method With Dielectric Regions. The integral-equation method can be extended to deal with situations for which the bounded region consists of two or more dielectrics, e.g. as one might expect to find in some semiconductor devices. In Fig. 4-14 we have sketched a two-dielectric medium bounded by surface $S = S_1 + S_2$, and we define S_{12} to be the interface between the two dielectrics with permittivities ϵ_1 and ϵ_2, respectively. We intend to evaluate $\Phi(\vec{r}_P)$ at point P in the ϵ_1 region. The electrostatic field at P is determined uniquely by the value of $\hat{n}_Q \cdot \vec{E}(\vec{r}_Q)$ at all boundary points $\vec{r}_Q$ on S and S_{12}. The dotted line in Fig. 4-14 shows $S_1 + S_2 + S_{12}$ as the limit of one continuous bounding surface. Let $\omega_s(\vec{r}_Q) = \epsilon_0 \hat{n}_Q \cdot \vec{E}(\vec{r}_Q)$, where the surface-normal unit vector $\hat{n}_Q$ points *into* the dielectric regions. Then we may write (Stratton 1941),

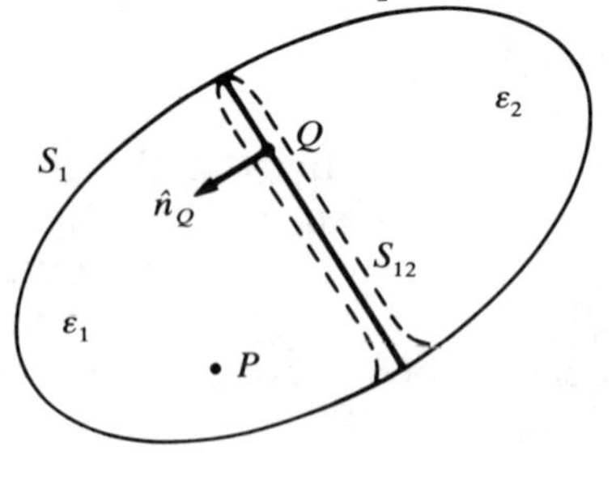

Fig. 4-14. Treatment of two uniform regions with differing dielectric permittivities bounded by a surface with known potential distribution. Note apparent surface doubling of interfaces.

$$\Phi(\vec{r}_P) = \frac{1}{4\pi\epsilon_0}\left[\int_S\int dS_Q \frac{\omega_s(\vec{r}_Q)}{R_{PQ}} + \int_{S_{12}}\int dS_Q \frac{[\omega_s(\vec{r}_Q) - \bar{\omega}_s(\vec{r}_Q)]}{R_{PQ}}\right], \tag{4.37}$$

where $\bar{\omega}_s(\vec{r}_Q) = \epsilon_0 \hat{n}_Q \cdot \vec{E}_2$ for the contribution from the ϵ_2 side of S_{12} so that the minus sign signals that we have adopted the convention of having $\hat{n}_Q$ point *into* ϵ_1 even for the surface vector on the ϵ_2 side. With these definitions, the boundary condition for normal-$\vec{E}$ components at the interface S_{12} becomes

$$\epsilon_1 \omega_s(\vec{r}_Q) = \epsilon_2 \bar{\omega}_s(\vec{r}_Q). \tag{4.38}$$

Therefore, Eq. (4.37) can be rewritten as

$$\Phi(\vec{r}_P) = \frac{1}{4\pi\epsilon_0}\left[\int_S\int dS_Q \frac{\omega_s(\vec{r}_Q)}{R_{PQ}} + \left(1 - \frac{\epsilon_1}{\epsilon_2}\right)\int_{S_{12}}\int dS_Q \frac{\omega_s(\vec{r}_Q)}{R_{PQ}}\right]. \tag{4.39}$$

We discretize surface S $(= S_1 + S_2)$ into N $(= N_1 + N_2)$ elements, and likewise S_{12} into M elements. Hence, there are $N + M$ discrete values of surface-charge density $\omega_s(\vec{r}_Q)$ that need to be calculated, but the problem is that the boundary potential $\Phi(\vec{r}_s)$ is known only on S and not on S_{12}. Therefore, it appears as if we have only N linear inhomogeneous equations in $N + M$ variables $\omega_s(\vec{r}_Q)$. We must find M more equations or eliminate M variables!

The solution to this dilemma is found by application of Eq. (4.39) to the right-hand side of the following equation for M points Q on the ϵ_1 side of the S_{12} interface (see Fig. 4-15 for clarification):

$$\omega_s(\vec{r}_Q) = \lim_{h\to 0}[\epsilon_0 \hat{n}_Q \cdot \vec{E}_1(\vec{r}_Q)] = \lim_{h\to 0}[-\epsilon_0 \hat{n}_Q \cdot \vec{\nabla}_Q \Phi(\vec{r}_Q)]. \tag{4.40}$$

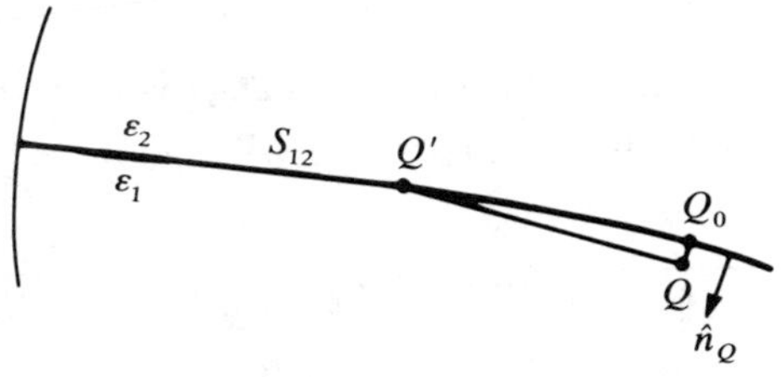

Fig. 4-15. The limiting procedure required for evaluation of the integrals in Eqs. (4.42) and (4.43).

In Fig. 4-15, we define $QQ_0 = h$ so that application of Eq. (4.40) to (4.39) at point Q will give

$$\omega_s(\vec{r}_Q) = \lim_{h \to 0} \frac{1}{4\pi\epsilon_0} \left\{ \int_S \int dS_{Q'}\, \omega_s(\vec{r}_{Q'}) \hat{n}_Q \cdot \vec{\nabla}_Q \frac{1}{R_{QQ'}} + \left(1 - \frac{\epsilon_1}{\epsilon_2}\right) \int_{S_{12}} \int dS_{Q'}\, \omega_s(\vec{r}_{Q'}) \hat{n}_Q \cdot \vec{\nabla}_Q \frac{1}{R_{QQ'}} \right\}. \tag{4.41}$$

We may bring the normal-gradient operator $\hat{n}_Q \cdot \vec{\nabla}_Q$ to the left outside of the integrations in the first integral. The second integral in Eq. (4.41) requires some further preparation. Specifically, we note that

$$\hat{n}_Q \cdot \vec{\nabla}_Q \frac{1}{R_{QQ'}} = -\frac{\hat{n}_Q \cdot \hat{R}_{QQ'}}{R^2_{QQ'}} = \frac{h}{(R^2_{Q_0Q'} + h^2)^{3/2}} \tag{4.42}$$

We must be careful about the integral in the immediate neighborhood of point Q_0 when taking the limit $h \to 0$. Let us abbreviate $R_{Q_0Q'} = r$. We are dealing with infinitesimal r in (4.42) as $h \to 0$, in which case the surface element $dS_Q \approx 2\pi r\, dr$ will be an excellent approximation for its area for nonzero r in the interface, and therefore

$$\int_{S_{12}} \int dS_{Q'}\, \omega_s(\vec{r}_{Q'}) \hat{n}_Q \cdot \vec{\nabla}_Q \frac{1}{R_{QQ'}} \approx \omega_s(\vec{r}_{Q_0}) 2\pi \lim_{h \to 0} \int_0^{r_0} dr \frac{hr}{(r^2 + h^2)^{3/2}} \approx 2\pi \omega_s(\vec{r}_{Q_0}) . \tag{4.43}$$

The upper bound r_0 is an infinitesimal but nonzero value so that $r_0/h \to \infty$ as $h \to 0$. As a result, Eq. (4.41) can then be written as

$$\omega_s(\vec{r}_Q) = \hat{n}_Q \cdot \vec{\nabla}_Q \left[\frac{1}{4\pi\epsilon_0} \int_S \int dS_{Q'} \frac{\omega_s(\vec{r}_{Q'})}{R_{QQ'}} \right] + \frac{1}{2}\left(1 - \frac{\epsilon_1}{\epsilon_2}\right) \omega_s(\vec{r}_Q) , \tag{4.44}$$

after we have gone to the limit $h \to 0$ (so that Q is then on the interface). Combining the terms with $\omega_s(\vec{r}_Q)$, we obtain

$$\omega_s(\vec{r}_Q) = \frac{2}{1 + (\epsilon_1/\epsilon_2)} \hat{n}_Q \cdot \vec{\nabla}_Q \left[\frac{1}{4\pi\epsilon_0} \int_{S_1} \int dS_{Q'} \frac{\omega_s(\vec{r}_{Q'})}{R_{QQ'}} \right]. \tag{4.45}$$

When this equation is combined with Eq. (4.39), we obtain for any interior point in the ϵ_1 region,

$$\Phi(\vec{r}_P) = \Phi_0(\vec{r}_P) + \left(\frac{1-(\epsilon_1/\epsilon_2)}{1+(\epsilon_1/\epsilon_2)}\right)\frac{1}{2\pi}\int_{S_{12}}\int dS_Q \,\frac{\hat{n}_Q\cdot\vec{\nabla}_Q\Phi_0(\vec{r}_Q)}{R_{PQ}}\,,$$

$$\Phi_0(\vec{r}_P) \equiv \frac{1}{4\pi\epsilon_0}\int_S\int dS_Q \,\frac{\omega_s(\vec{r}_Q)}{R_{PQ}}\,. \tag{4.46}$$

Now we require only N points Q on S, and M on S_{12} to determine the discretized source densities $\omega_s(\vec{r}_Q)$ from Eq. (4.46). The additional M discretizations along S_{12} allow us to evaluate all $\omega_s(\vec{r}_Q)$ from the $N+M$ linear inhomogeneous equations that result from the discretization of Eq. (4.46).

If there are more than two dielectrics, then there will be a sum of more terms in Eq. (4.46) with factor $[1-(\epsilon_i/\epsilon_j)]/[1+(\epsilon_i/\epsilon_j)]$ for each interface S_{ij}. The unit vector $\hat{n}_Q$ in each is chosen consistently to point into medium ϵ_i.

4.1.5.4 Finite-Mesh Relaxation versus Integral-Equation Methods. Consider a free-space situation with a number of constant-voltage plates determining the potentials and fields in the regions between them. What considerations go into deciding whether to use a finite-mesh relaxation or an integral-equation method? This section offers the reader some aids in making a rational decision.

Relaxation methods require knowledge of the potential Φ at all points $\vec{r}_s$ on a *closed* boundary surface S. If the surface is not closed, then it may be difficult to use this method, unless one can estimate $\Phi(\vec{r}_s)$ on some surface(s) interpolated between two (or more) conductors. For example, two semi-infinite cylinders of the same diameter, placed end-to-end with a relatively small gap between them (see Fig. 4-16) are "open" with respect to their interiors in the vicinity of the gap. Nevertheless, if the gap is less than the cylinder radius, it can be demonstrated that a linear interpolation of potential between V_1 on one cylinder and V_2 on the other for the extension of the cylindrical surfaces in the gap is an excellent approximation. The boundary surface has then been closed, and $V(\vec{r}_s)$ is known at all its points.

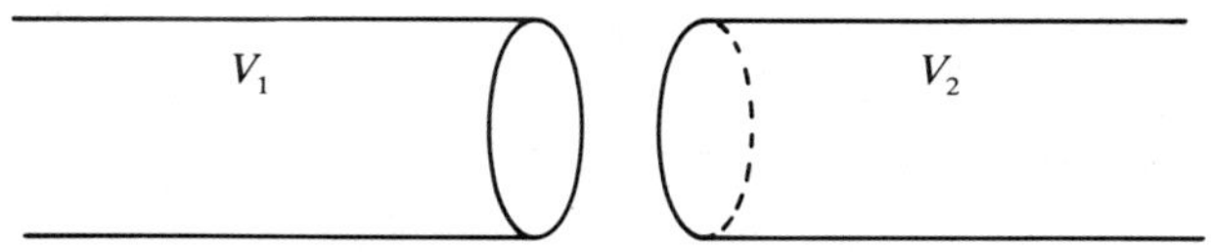

Fig. 4-16. Two circular metallic co-axial cylinders of equal radius with a small gap between them.

Relaxation methods have special advantages when the total number of mesh points $N = N_1 N_2 N_3$ (in three dimensions) is not excessive so that an $N \times N$ matrix can be inverted directly or through an iterative procedure. The inversion yields values of $\Phi(\vec{r})$ at all mesh points $\vec{r} = \vec{r}_n$, and fields can then be obtained numerically from differences between neighboring potential values.

However, situations with nonclosed boundaries easily develop for which the number of mesh points would be excessive if the boundary is closed artificially. Even when $N_1 = N_2 = N_3 = 100$, we would have to invert a $10^6 \times 10^6$ matrix—a formidable task. In situations that would otherwise call for so many mesh points it may seem advantageous to use an integral-equation method, because such situations usually have only several conductor surfaces, the sum of which need to be discretized in two dimensions into N surface elements. The chances are high that this N is much less than the number of mesh points in three dimensions. The inversion of the $N \times N$ matrix for surface charge densities $\omega_s(\vec{r}_s)$ would then be less formidable. However, it is still necessary, according to Eq. (4.32), to evaluate N integrals after that to obtain $\Phi(\vec{r})$ at an interior point. This method may also become very expensive if the potential at many interior points is required.

Relaxation methods adapt better to two-dimensional situations for which $N = N_1 N_2$ can be kept reasonably small. One common method for increasing the speed of calculation is to start in a region with a relatively coarse grid, and then to define a subregion in which greater accuracy is desired. One can then interpolate potential values on the boundary surface of the subregion and then continue a relaxation on a finer mesh inside the subregion.

Another compromise of use for a large region with nonclosed boundary is first to employ an integral-equation method to determine $\Phi(\vec{r})$ on a two-dimensional surface enclosing a volume of interest, and then to use a relaxation method with these boundary potentials to determine interior potentials. The advantage occurs when the volume of interest requires less points than the number of surface discretizations, and when one needs to compute interior potentials at constantly changing interior points. However, the trade-off usually requires careful thought before deciding upon an optimal procedure.

4.2 MAGNETOSTATIC FIELDS

Calculation of magnetostatic fields is often (but not always) similar to that of electrostatic fields, and this section is devoted to a discussion of those aspects that one needs to pay attention to most.

4.2.1 The Simplest Analytical Solutions

With reference to Section 4.1.1, and also specifically to standard textbooks such as those cited there, the simplest cases are as follows.

1. Magnetic field around a circular infinite straight wire with constant radius ρ_0 (see Fig. 4-17). Ampère's law yields

$$\begin{aligned}\vec{H}(\rho) &= -\frac{\rho}{2\pi\rho_0^2}\hat{\rho} \times \vec{I} \quad \text{for } \rho \leqslant \rho_0 , \\ &= -\frac{1}{2\pi\rho}\hat{\rho} \times \vec{I} \quad \text{for } \rho > \rho_0 , \end{aligned} \tag{4.47}$$

given that $I = \pi r_0^2 J$ is the uniform current along the wire (vector indicates direction).

2. An infinite circular solenoid with n windings of current I per unit length (see Fig. 4-18. In this case we easily observe that

$$\begin{aligned}\vec{H} &= \mu_0 n\hat{\rho} \times \vec{I} \quad \text{for } \rho < \rho_0 , \\ &= 0 \quad \text{for } \rho > \rho_0 . \end{aligned} \tag{4.48}$$

The solution is similar for a solenoid filled with a uniform magnetic material provided μ_0 is replaced by the permeability μ of that material.

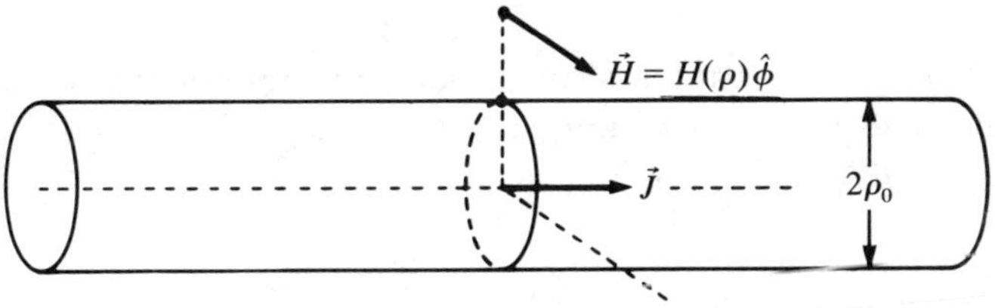

Fig. 4-17. An circular infinite straight wire shown as illustration of Ampère's law.

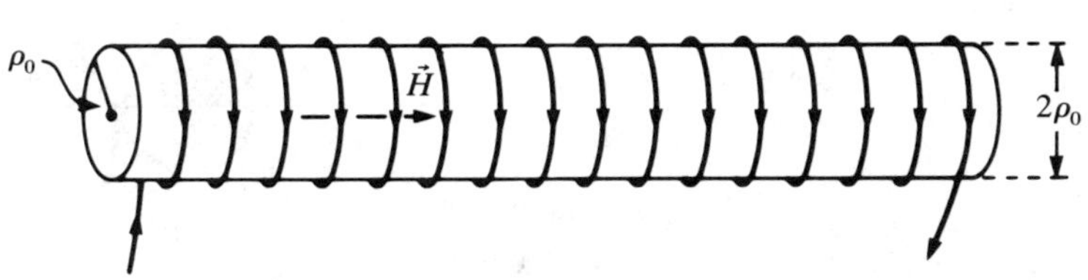

Fig. 4-18. A solenoid consisting of current windings around a circular cylindrical base.

Extensions of these solutions can be obtained by means of Ampère's law for several other simple cases, but we will restrict ourselves to just the above.

4.2.2 The Scalar Magnetic Potential

A large class of applications requires knowledge of the magnetic field $\vec{H}$ in a uniform magnetic material. The magnetostatic equations are

$$\vec{\nabla} \times \vec{H} = \vec{J}\,, \qquad \vec{\nabla} \cdot \vec{H} = 0\,. \tag{4.49}$$

If solutions for $\vec{H}$ are required in regions outside the sources, i.e., where $\vec{J} = 0$, then the curl-H equation may be replaced by defining a *scalar magnetic potential* Φ_m by means of

$$\vec{H} = -\vec{\nabla}\Phi_m\,, \tag{4.50}$$

and then the second of Eqs. (4.49) results in a Laplace equation for Φ_m outside the source regions. In this case, Neumann conditions (specification of $\hat{n} \cdot \vec{\nabla}\Phi_m = -\hat{n} \cdot \vec{H}$) will be given usually on the bounding surface of the region of interest. A problem here is that Φ_m is multivalued because Eq. (4.50) yields

$$\Phi_m(\vec{r}_P) - \Phi_m(\vec{r}_Q) = -\int_Q^P \vec{dl} \cdot H\,. \tag{4.51}$$

Figure 4-19 shows a line integral from point Q to point P that encircles a line-current density $I(\vec{r}) = I\hat{z}\delta(\vec{\rho})$ once entirely. The contour can be deformed into that of Fig. 4-20, and then

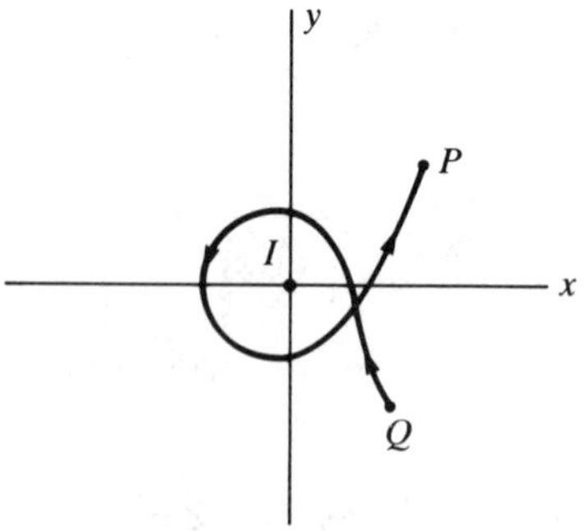

Fig. 4-19. A line integral that figures in Eq. (4.51) for the scalar magnetic potential before deformation.

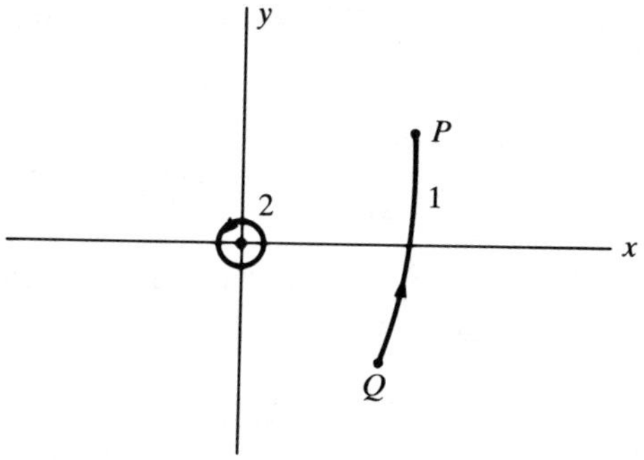

Fig. 4-20. Deformation of the line integral in Eq. (4.51) to account for multivaluedness.

$$\Phi_m(\vec{r}_P) - \Phi_m(\vec{r}_Q) = -\int_Q^P (\vec{dl} \cdot \vec{H})_1 - \int_Q^P (\vec{dl} \cdot \vec{H})_2$$

$$= -\int_Q^P (\vec{dl} \cdot \vec{H})_1 - I . \qquad (4.52)$$

The infinitesimal contour 2 yields the enclosed current (the sign is determined by the right-hand rule), and the potential $\Phi_m(\vec{r}_P)$ therefore depends on whether the line integral from point Q to P encircles currents or not. Obviously I will be replaced by nI in Eq. (4.52) if the current is encircled n times.

4.2.3 Vector Potential Methods

4.2.3.1 Currents in Vacuum. In a situation in which current distributions can be approximated by infinitesimally thin wires or by two-dimensional sheets of current bounding a region where one wishes to evaluate $\vec{H}(\vec{r})$, it is often sufficient to use the *vector magnetic potential* defined by $\vec{H} = \vec{\nabla} \times \vec{A}$ with $\vec{\nabla} \cdot \vec{A} = 0$. The alternate definition of $\vec{A}$ via $\vec{B} = \vec{\nabla} \times \vec{A}$ is used in many textbooks but the difference is inessential. From the magnetostatic Maxwell equations (4.49) and the vanishing of div $\vec{A}$, it is easily found that

$$\Delta\vec{A} = -\vec{J} , \qquad \vec{A}_P = \frac{1}{4\pi} \iiint dv_Q \, \frac{\vec{J}_Q}{R_{PQ}} . \qquad (4.53)$$

Alternative forms for sheet currents and idealized wire-current distributions are

$$\vec{A}_P = \frac{1}{4\pi}\int\int dS_Q \frac{\vec{K}_Q}{R_{PQ}}, \qquad \vec{A}_P = \frac{I}{4\pi}\int \vec{dl}_Q \frac{1}{R_{PQ}}, \tag{4.54}$$

given that $\vec{K}_Q$ (in A/m) is a two-dimensional sheet-current density.

Magnetic fields are often produced by coil-wound (or otherwise wound) distributions of current-carrying wire that, to first approximation, may be considered as sheet currents. Such current distributions produce vector magnetic potentials as given in Eq. (4.54). Each vector component can be handled in principle by integral-equation methods, as put forth in Section 4.1.5.2. The $\vec{H}$ field can then be obtained numerically by applying the difference-equation form of $\vec{H} = \vec{\nabla} \times \vec{A}$.

4.2.3.2 Inclusion of Magnetic Materials. Often, as in magnets and deflection yokes in TV kinescopes, current windings are around materials with permeability μ appreciably different from the free-space permeability μ_0. Infinitesimally thin wires, of course, can have different μ but these can be modeled as line or sheet currents. In applications with high-permeability materials, we need to take into account explicitly the presence of $\mu \neq \mu_0$ regions. The boundary conditions at material-vacuum interfaces, as depicted in Fig. 4-21, are

$$\hat{n} \cdot [\vec{H}_0 - \mu_r \vec{H}] = 0, \qquad \vec{n} \times [\vec{H}_0 - \vec{H}] = \vec{K}. \tag{4.55}$$

Here, $\mu_r = \mu/\mu_0$ is the relative permeability, and the subscript "0" indicates the field on the free-space side of the interface. In the absence of sheet-current sources ($\vec{K} = 0$), and for large values of μ_r, the normal component of $\vec{H}$ in the material will be much less than that in free space

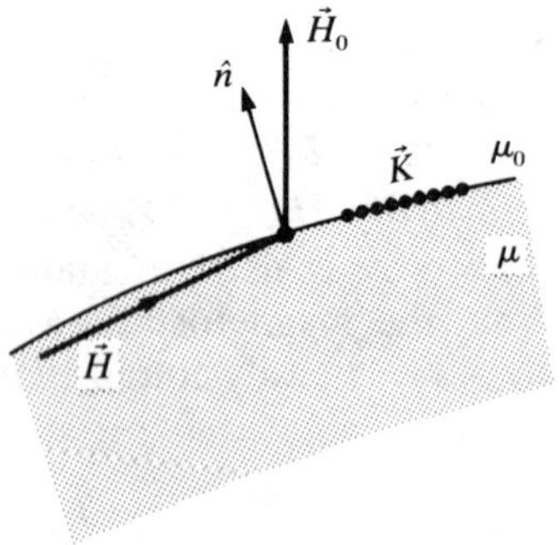

Fig. 4-21. Material-vacuum interfaces with surface-current distributions.

at the interface, and either $\vec{H}_0$ is almost perpendicular or $\vec{H}$ is almost parallel to the interface. One method of obtaining the $\vec{H}$ field from a given sheet-current density $\vec{K}(\vec{r})$ is to consider a superposition $\vec{H}(\vec{r}) = \vec{H}^{(0)}(\vec{r}) + \vec{H}^{(m)}(\vec{r})$ such that $\vec{H}^{(0)}(\vec{r})$ is the field produced *everywhere* by the source sheet current with density $\vec{K}(\vec{r})$ in the *absence* of high-μ material. In that case,

$$H_P^{(0)} = \vec{\nabla} \times \vec{A}_P , \qquad \vec{A}_P = \frac{1}{4\pi} \int\int dS_Q \, \frac{\vec{K}_Q}{R_{PQ}} , \tag{4.56}$$

and methods analogous to those of Section 4.1.5.2 can be applied, in principle, to the solution of these equations. Because $\vec{H}(\vec{r})$ and $\vec{H}^{(0)}(\vec{r})$ have identical source distributions, it follows that $\vec{H}^{(m)}(\vec{r})$ is irrotational (has zero curl). Consequently, a scalar potential $\Phi^{(m)}(\vec{r})$ can be introduced,

$$H_P^{(m)} = -\vec{\nabla}\Phi_P^{(m)} , \tag{4.57}$$

which obeys Laplace's equation everywhere except at the interface between free space and the μ material.

Consider a point P in free space, and let the surface S consist of the interfaces with the μ material and S_∞ (the infinitely distant "surface"). Application of a Green's theorem from classical potential theory (Stratton 1941) yields

$$\Phi_P^{(m)} = -\frac{1}{4\pi} \int_S \int \overrightarrow{dS}_Q \cdot \left[\frac{1}{R_{PQ}} \, \vec{\nabla}_Q \Phi_Q^{(m)} - \Phi_Q^{(m)} \vec{\nabla}_Q \, \frac{1}{R_{PQ}} \right], \tag{4.58}$$

with $\overrightarrow{dS}_Q = dS_Q \, \hat{n}$ such that $\hat{n}$ is a unit vector normal to the surface, pointing *into* the free-space region (out of the μ material at the interfaces). This is rewritten as

$$\begin{aligned} \Phi_P^{(m)} &= \frac{1}{4\pi} \int_S \int dS_Q \, \frac{\hat{n} \cdot \vec{H}_Q^{(m)}}{R_{PQ}} + \delta\Phi_P^{(m)} , \\ \delta\Phi_P^{(m)} &= \frac{1}{4\pi} \int_S \int dS_Q \, \Phi_Q^{(m)} \hat{n} \cdot \vec{\nabla}_Q \, \frac{1}{R_{PQ}} . \end{aligned} \tag{4.59}$$

In the next few lines, $\delta\Phi_P^{(m)}$ will be somewhat transformed. First replace $\hat{n}$ at point Q by $\hat{n}' = -\hat{n}$ so that $\overrightarrow{dS}_Q$ becomes $-\overrightarrow{dS}'_Q$ with $\overrightarrow{dS}'_Q$ pointing *into* the μ material. At this stage we may safely interpret S as the sum of all interfaces with μ material. Now we have

$$\begin{aligned}\delta\Phi_P^{(m)} &= -\frac{1}{4\pi}\int_S\int \overrightarrow{dS'_Q}\cdot\Phi_Q^{(m)}\vec{\nabla}_Q\,\frac{1}{R_{PQ}}\\ &= -\frac{1}{4\pi}\int\int_{v_m}\int dv_Q\,\vec{\nabla}_Q\cdot\left[\Phi_Q^{(m)}\vec{\nabla}_Q\,\frac{1}{R_{PQ}}\right].\end{aligned} \tag{4.60}$$

Here, v_m is the total volume occupied by the μ material, and Gauss's divergence theorem has been applied. Note that point P lies *outside* v_m. Therefore, for all values of $\vec{r}_Q \in v_m$ we may interpret the Laplacian of $1/R_{PQ}$ with respect to the point Q to be zero, so that Eq. (4.60) reduces to

$$\begin{aligned}\delta\Phi_P^{(m)} &= -\frac{1}{4\pi}\int\int_{v_m}\int dv_Q\,\vec{\nabla}_Q\Phi_Q^{(m)}\cdot\vec{\nabla}_Q\,\frac{1}{R_{PQ}}\,.\\ &= \frac{1}{4\pi}\int\int_{v_m}\int dv_Q\,\vec{H}_Q^{(m)}\cdot\vec{\nabla}_Q\,\frac{1}{R_{PQ}} = \frac{1}{4\pi}\int\int_{v_m}\int dv_Q\,\vec{\nabla}_Q\cdot\left[\frac{\vec{H}_Q^{(m)}}{R_{PQ}}\right].\end{aligned} \tag{4.61}$$

The last equality has made use of the fact that the divergence of $\vec{H}_Q^{(m)}$ is zero everywhere in v_m (except on interfaces). We then apply Gauss's divergence theorem once again to obtain

$$\delta\Phi_P^{(m)} = \frac{1}{4\pi}\int_S\int d\vec{S}'_Q\cdot\frac{\vec{H}_Q^{(m)}}{R_{PQ}} = -\frac{1}{4\pi}\int_S\int dS_Q\,\frac{\hat{n}\cdot\vec{H}_Q^{(m)}}{R_{PQ}}\,. \tag{4.62}$$

It should also be stated that $\vec{H}_Q^{(m)}$ is on the μ-material side of the interface. Consequently, the use of Eq. (4.62) in (4.59) yields

$$\Phi_P^{(m)} = \frac{1}{4\pi}\int_S\int dS_Q\,\frac{1}{R_{PQ}}\,\{[\hat{n}\cdot\vec{H}_Q^{(m)}]_0 - [\hat{n}\cdot\vec{H}_Q^{(m)}]_m\}\,. \tag{4.63}$$

The subscripts on the square brackets denote the free-space side (0) and the material side (m) of the interfaces. Now use

$$\begin{aligned}[\hat{n}\cdot\vec{H}_Q^{(m)}]_0 &= [\hat{n}\cdot\vec{H}_Q]_0 - \hat{n}\cdot\vec{H}_Q^{(0)}\,,\\ [\hat{n}\cdot\vec{H}_Q^{(m)}]_m &= [\hat{n}\cdot\vec{H}_Q]_m - \hat{n}\cdot\vec{H}_Q^{(0)}\,.\end{aligned} \tag{4.64}$$

Making use of the fact that $\vec{H}_Q^{(0)}$ is continuous across all vacuum-material interfaces, we subtract both sides of Eq. (4.64) from each other and replace $[\hat{n}\cdot\vec{H}_Q]_m$ by $\mu_r^{-1}[\hat{n}\cdot\vec{H}_Q]_0$ to obtain

$$[\hat{n} \cdot \vec{H}_Q^{(m)}]_0 - [\hat{n} \cdot \vec{H}_Q^{(m)}]_m = [\hat{n} \cdot \vec{H}_Q]_0 - \frac{1}{\mu_r} [\hat{n} \cdot \vec{H}_Q]_0 \,. \tag{4.65}$$

Therefore, Eqs. (4.64) and (4.65) combine to modify Eq. (4.63) into

$$\Phi_P^{(m)} = \frac{1 - \mu_r^{-1}}{4\pi} \int_S \int dS_Q \, \frac{[\hat{n} \cdot \vec{H}_Q]_0}{R_{PQ}} \,. \tag{4.66}$$

It is gratifying to see that the material-induced scalar potential vanishes when $\mu_r = 1$. With Eq. (4.57), we then obtain

$$\vec{H}_P^{(m)} = \frac{1 - \mu_r^{-1}}{4\pi} \int_S \int dS_Q \, \frac{[\hat{n} \cdot \vec{H}_Q]_0}{R_{PQ}^2} \hat{R}_{PQ} \,, \tag{4.67}$$

where $\hat{R}_{PQ}$ is a unit vector pointing from surface location Q to observation point P. The solution $\vec{H}_P$ is thus the sum of the expressions given in Eqs. (4.56) and (4.67). It can be obtained in numerical fashion by an integral-equation method which is outlined below.

The normal component $\hat{n} \cdot \vec{H}$ at any point on the interface is essentially equivalent to a surface-charge density, call it ω. Mutiply Eq. (4.67) by $\hat{n}_P$ on both sides to obtain

$$\begin{aligned} \omega_P^{(m)} &= -\frac{1 - \mu_r^{-1}}{4\pi} \int_S \int dS_Q \, [\omega_Q]_0 \hat{n}_P \cdot \vec{\nabla}_P \frac{1}{R_{PQ}} \\ &= -\frac{1 - \mu_r^{-1}}{4\pi} \int_S \int dS_Q \, [\omega_Q^{(0)} + \omega_Q^{(m)}] \hat{n}_P \cdot \vec{\nabla}_P \frac{1}{R_{PQ}} \end{aligned} \tag{4.68}$$

in the understanding that $\omega_Q^{(m)}$ pertains to the free-space side of the interface. After discretization into N surface patches S_j, and in the limiting case in which P approaches the center of one of those surface patches, we obtain for $1 \leq i \leq N$,

$$\begin{aligned} \omega_i^{(m)} &= (1 - \mu_r^{-1}) \sum_{j=1}^{N} c_{ij} [\omega_j^{(0)} + \omega_j^{(m)}] \,, \\ c_{ij} &\equiv -\frac{1}{4\pi} \int_{S_j} \int dS' \, \hat{n}_i \cdot \vec{\nabla} \frac{1}{|\vec{r}_i - \vec{r}'|} \,. \end{aligned} \tag{4.69}$$

The quantities $\omega_j^{(0)} = \hat{n}_j \cdot \vec{H}_j^{(0)}$ are obtained from a discretization of Eq. (4.56), or from a direct solution of those equations at the N interface points. When all $\omega_i^{(m)}$ and $\omega_i^{(0)}$ have thus been obtained, Eqs. (4.67) and

(4.56) will lead to $\vec{H}(\vec{r})$ at arbitrary free-space locations by means of a similar discretization of the integrands.

This, then, ends our survey of evaluation methods for determining electrostatic and magnetostatic fields. In the next chapter we turn our attention towards lens calculations via optical analogs.

REFERENCES

Bonjour, P. (1980) *Applied Charged Particle Optics* (Ed. A. Septier), Vol I (*Advances in Electronics and Electron Physics*, Suppl. 13A), Academic Press, New York.

Cheng D.K. (1989) *Field and Wave Electromagnetics*, 2nd edn., Addison-Wesley, New York.

de Wolf, D.A. (1978) *Proc. IEEE*, **66**, 85–86

de Wolf, D.A. (1981) *Proc. IEEE*, **69**, 123–124.

Forsyth, G.E. and W.R. Wasov (1960) *Finite-Difference Methods for Partial Difference Equations*, John Wiley, New York.

Harrington, R.F. (1968) *Field Computation by Moment Methods*, Macmillan, New York.

Hayt, W.H. (1989) *Engineering Electromagnetics*, 5th edn., McGraw-Hill, New York.

Kraus, J.D. (1984) *Electromagnetics*, 3rd edn., McGraw-Hill, New York.

Landau, L.D. and E.M. Lipschitz (1960) *Electrodynamics of Continuous Media*, Pergamon, London.

Neff Jr., H.P. (1987) *Basic Electromagnetic Fields*, 2nd edn., Harper & Row, Cambridge.

Smythe W.R. (1969) *Static and Dynamic Electricity*, 3rd edn., McGraw-Hill, New York.

Stratton, J.A. (1941) *Electromagnetic Theory*, McGraw-Hill, New York.

Varga, R.S. (1971) *Matrix Iterative Analysis*, 2nd edn., Prentice-Hall, Englewood Cliffs, NJ.

Weber, C. (1967) *Focussing of Charged Particles* (Ed. A. Septier), Vol I, Academic Press, New York.

CHAPTER 5

PARAXIAL (GAUSSIAN) OPTICS

Much insight into the behavior of electron beams in "lenses" formed by electromagnetic fields can be gained by studying the approximation known as *paraxial* (or *Gaussian*) optics. Paraxial optics are the result of approximations in the trajectory equations that result from exploiting the "smallness" of $\vec{\rho}(z)$ and $\vec{\rho}\,'(z)$. The meaning of "smallness" needs yet to be defined; however, the paraxial approximation is well known in wave optics (Goodman 1986), where it leads to a replacement of the scalar wave equation by a simpler parabolic equation when it is known that all initially planar wave fronts deviate from their original direction only by very small angles. There are more criteria, but that is the main one. The paraxial approximation is somewhat more restrictive in electron optics because paraxial electron beams, unlike beams in wave optics, may not become wide. All this will be clarified, but first let us note that the criterion for the ray tangent in electron optics is

$$|\vec{\rho}\,'|^2 \ll 1 \,. \tag{5.1}$$

Referring to Chapter 3, we obtain an approximation for the equation of motion of a charged particle in a purely electrostatic field (at nonrelativistic velocities):

$$\vec{\rho}\,''(z) \approx \frac{1}{2\Phi_a}\left(\vec{\nabla}_T\Phi - \vec{\rho}\,'\,\frac{\partial \Phi}{\partial z}\right), \tag{5.2}$$

given that $\vec{\nabla}_T = \hat{x}\partial/\partial x + \hat{y}\partial/\partial y$ in Cartesian coordinates. Define polar coordinates ρ, ϕ so that

$$\begin{aligned} x &= \rho \cos \phi \,, \\ x' &= \rho' \cos \phi - \rho\phi' \sin \phi \,, \\ x'' &= [\rho'' - \rho(\phi')^2] \cos \phi - [2\rho'\phi' + \rho\phi''] \sin \phi \,, \end{aligned} \tag{5.3}$$

and similar expressions for $y = \rho \sin \phi$, y', and y''. By using the x and y components of $\vec{\rho}''$ in Eq. (5.2) to form the combination $x'' \cos \phi + y'' \sin \phi$ and $x'' \cos \phi - y'' \sin \phi$, one obtains with Eq. (5.3),

$$\begin{aligned} \rho'' - \rho(\phi')^2 &\approx \frac{1}{2\Phi_a}\left(\frac{\partial \Phi}{\partial \rho} - \rho' \frac{\partial \Phi}{\partial z}\right), \\ \phi'' + 2\phi' \frac{\rho'}{\rho} &\approx \frac{1}{2\Phi_a}\left(\frac{1}{\rho^2}\frac{\partial \Phi}{\partial \phi} - \phi' \frac{\partial \Phi}{\partial z}\right). \end{aligned} \tag{5.4}$$

These equations are identical to (5.2), but they are in polar form. Both have one of the conditions for paraxial equations built into them; but we have yet to require that the radial trajectory parameter ρ be small (and to define what "small" implies).

5.1 ELECTROSTATIC TRAJECTORY EQUATIONS—CYLINDRICAL SYMMETRY

An important and relatively simple case is one in which all the fields are *axisymmetric*, i.e., if the fields are electrostatic (as we assume here) then $\Phi(\rho, z)$ is a function only of the cylindrical coordinates ρ and z, but not of the azimuthal angle ϕ. In that case, Eqs. (5.4) simplify somewhat to

$$\begin{aligned} \rho'' - \rho(\phi')^2 &\approx \frac{1}{2\Phi_a}\left(\frac{\partial \Phi}{\partial \rho} - \rho' \frac{\partial \Phi}{\partial z}\right), \\ \phi'' + 2\phi' \frac{\rho'}{\rho} &\approx -\frac{\phi'}{2\Phi_a}\frac{\partial \Phi}{\partial z} \quad \text{or} \quad \frac{d}{dz}[\sqrt{\Phi_a}\rho^2\phi'] \approx 0 \,. \end{aligned} \tag{5.5}$$

Note that the second equation does not have a $\partial\Phi/\partial\phi$ term; in other respects (5.5) is unchanged from Eqs. (5.4).

Referring to Eq. (3.70), the general expansion for axisymmetric potential, we now make the second, more important paraxial approximation,

$$\Phi(\rho, z) \approx \Phi_0(z) - \tfrac{1}{4}\rho^2\Phi_0''(z) , \tag{5.6}$$

where $\Phi_0(z)$ is a convenient abbreviation for the axial potential $\Phi(0, z)$. This approximation requires that the missing terms are negligible and specifically it is necessary (although possibly not sufficient) that

$$|\tfrac{1}{4}\rho^2\Phi_0''(z)| \ll |\Phi_0(z)| , \tag{5.7a}$$

or, equivalently

$$\rho^2 \ll 4\left|\frac{\Phi_0(z)}{\Phi_0''(z)}\right| . \tag{5.7b}$$

This criterion obviously fails at turning points where it must be replaced by another fairly obvious criterion, and one must also be assured that the the higher powers of ρ^2 vanish sufficiently rapidly. Clearly, Eq. (5.6) requires small ρ, and the question "how small?" is answered by investigating the inequalities (5.7) and similar ones involving higher-order terms. The paraxial approximation (5.6) implies

$$\frac{\partial\Phi}{\partial\rho} \approx -\frac{1}{2}\Phi_0''\rho , \qquad \frac{\partial\Phi}{\partial z} \approx \Phi_0' . \tag{5.8}$$

Equations (5.5) then become

$$\rho'' - \rho(\phi')^2 \approx -\frac{1}{4}\frac{\Phi_0''}{\Phi_{a0}}\rho - \frac{1}{2}\frac{\Phi_0'}{\Phi_{a0}}\rho' ,$$
$$\frac{d}{dz}(\sqrt{\Phi_a}\rho^2\phi') \approx 0 , \tag{5.9}$$

given that $\Phi_{a0} = \Phi_a(0, z)$, the value of the accelerating potential on the z axis. Equations (5.9) are the result of the paraxial approximations for *arbitrary* paraxial electrons. It is often sufficient in electrostatic situations to look only at *meridional* electron trajectories, i.e., trajectories that lie entirely in a plane that contains the z axis. Electrons with initial velocity vector $\vec{v}$ that lies in a plane containing the z axis will give rise to meridional trajectories in axisymmetric electrostatic fields. Such electrons are called *meridional electrons*, and Eqs. (5.9) reduce for them to

$$\rho'' = -\frac{1}{2}\frac{\Phi_0'}{\Phi_{a0}}\rho' - \frac{1}{4}\frac{\Phi_0''}{\Phi_{a0}}\rho , \tag{5.10}$$

which is much easier to solve (at least formally) than Eq. (5.9). Before discussing this paraxial equation further, let us mention an interesting alternative obtained from it due to Picht (El-Kareh and El-Kareh 1970). Introduce the *reduced* radial cordinate $R(z) = [\Phi_{a0}(z)]^{-1/4}\rho(z)$. If this is used in Eq. (5.10) to eliminate $\rho(z)$, we obtain the equation

$$R''(z) = -\frac{3}{16}\left[\frac{\Phi_0'(z)}{\Phi_0(z)}\right]^2 R(z)\,. \tag{5.11}$$

There are certain situations in which it is advantageous to use Eq. (5.11) instead of (5.10), especially with respect to some elementary-lens geometries. For example, a general numerical advantage lies in the fact that Eq. (5.11) requires knowledge only of $(\Phi_0')^2$, whereas Eq. (5.10) requires Φ_0'' for quadrature. Unless extra care is taken, the numerical integration of an integrand with Φ_0'' can yield significant errors, especially if Φ_0'' is first positive and then negative, as would be the case with an axial potential profile that is monotonically rising from an initial to a final value. As the resulting integral is then a difference of two large parts (each considerably larger than the difference), significant numerical error may occur in using Eq. (5.10) if extra care is not taken. Other differences will be discussed in connection with the consequences of these paraxial equations for various electron lenses.

The paraxial equations lead to (real or virtual) focussing and imaging. That is easily understood from their linearity. If a certain trajectory $\rho(z)$ with initial value $\rho(0)$ and initial slope $\rho'(0)$ crosses the axis at $z = z_f$, i.e., $\rho(z_f) = 0$, then any second trajectory with $\bar{\rho}(z) = c\rho(z)$, and with initial conditions $\bar{\rho}(0) = c\rho(0)$ and $\bar{\rho}'(0) = \rho'(0)$, for any constant c, will also cross the axis at $z = z_f$. Principal electrons (i.e., those with zero initial slope) thus all cross the axis at the same point, the focal point. Electrons emanating from one point ($z = 0$) also cross the axis again at one point, the image point.

5.2 ELECTROSTATIC TRAJECTORY EQUATIONS—PLANAR SYMMETRY

The development of paraxial equations for planar symmetry, which implies that $\Phi(x, z)$ is independent of the Cartesian coordinate y and that $\Phi(-x, z) = \Phi(x, z)$, is quite similar to that in the previous section for axisymmetry. We start now from Eq. (5.2), which becomes

$$x'' \approx \frac{1}{2\Phi_a}\left(\frac{\partial\Phi}{\partial x} - x'\,\frac{\partial\Phi}{\partial z}\right). \tag{5.12}$$

We make the following approximations, using Eqs. (3.75) or (3.76),

$$\Phi_a \approx \Phi_{a0} = \Phi_a(0, z) ,$$
$$\frac{\partial \Phi}{\partial x} \approx -\Phi_0'' x , \tag{5.13}$$
$$\frac{\partial \Phi}{\partial z} \approx \frac{\partial \Phi_0}{\partial z} = \Phi_0' ,$$

in the understanding that the notation $\Phi_0(z) \equiv \Phi(0, z)$ implies the axial potential profile. These approximations then reduce Eq. (5.12) to

$$x'' = -\frac{1}{2}\frac{\Phi_0'}{\Phi_{a0}} x' - \frac{1}{2}\frac{\Phi_0''}{\Phi_{a0}} x , \tag{5.14}$$

which is the paraxial equation for planar symmetry of electrostatic fields. By means of very similar arguments as those of the previous section for axisymmetric lenses, it can be shown that Eq. (5.14) describes focussing and imaging in the $x = 0$ plane, hence is useful in analyzing slit lenses. Focussing and imaging do not imply points in this case; only that focus and image lie on a line perpendicular to any $y = \text{constant}$ plane. The line collapses to a point if there is no electron motion in the $\hat{y}$ direction. The approximations (5.14) have criteria for their validity similar to those of (5.7), e.g., $x^2 \ll 1$.

5.3 MAGNETOSTATIC TRAJECTORY EQUATIONS—CYLINDRICAL SYMMETRY

The development of nonrelativistic equations for electrons in magnetostatic fields is rather similar, and we shall only highlight it. The starting point is Eq. (3.15) which reduces to

$$x'' = -x'y'K_x + (1 + x'^2)K_y - y'K_z$$
$$y'' = -(1 + y'^2)K_x + x'y'K_y + x'K_z , \tag{5.15}$$

where $\vec{K} = e\vec{B}/mv_z$ is a vector that is related to the rotation radius mv/eB of an electron with transverse velocity v around a magnetic field line of strength B. In the case of cylindrical symmetry (axisymmetric $\vec{K}$) we can only allow K to depend upon ρ and z, not on ϕ.

We have already observed that Maxwell's equations, applied to source-free regions, lead to Laplace's equation for the scalar magnetic potential

$\Phi_m(\rho, z)$ (see Section 4.2.2), which can be written with (3.70) for the case of cylindrical symmetry as

$$\Phi_m(\rho, z) = J_0\left(\rho \frac{\partial}{\partial z}\right)\Phi_{m0}(z) , \tag{5.16}$$

given that $\Phi_{m0}(z)$ is an abbreviation for $\Phi_m(0, z)$. The *paraxial* approximation of (5.16) is,

$$\Phi_m(\rho, z) \approx \Phi_{m0}(z) - \tfrac{1}{4}\Phi''_{m0}(z)\rho^2 , \tag{5.17}$$

which requires that the missing terms be negligible and thus that the magnitude of permissible ρ is restricted. When this paraxial approximation is applied in cylindrical coordinates to the Maxwell equation $\vec{H} = -\vec{\nabla}\Phi_m$, we obtain through first order in ρ,

$$H_\rho \approx \tfrac{1}{2}\Phi''_{m0}\rho , \qquad H_\phi = 0 , \qquad H_z \approx -\Phi'_{m0} . \tag{5.18}$$

These equations establish that

$$H_\rho \approx -\tfrac{1}{2}H'_z\rho , \qquad K_\rho \approx -\tfrac{1}{2}K'_z\rho . \tag{5.19}$$

Now return to (5.15) and neglect terms quadratic in the (small) direction tangents, x', y',

$$x'' \approx K_y - y'K_z , \qquad y'' \approx -K_x + x'K_z . \tag{5.20}$$

We transform these expressions into similar ones for polar coordinates by using (5.3), and also by using (5.19) to transform K_x and K_y into

$$\begin{aligned} K_x &= K \cos\phi \approx -\tfrac{1}{2}K'_z\rho \cos\phi , \\ K_y &= K \sin\phi \approx -\tfrac{1}{2}K'_z\rho \sin\phi . \end{aligned} \tag{5.21}$$

Equations (5.20) then can be written as

$$\begin{aligned} \rho'' - \rho(\phi')^2 &= -K_z\rho\phi' , \\ \phi'' + 2\phi'\frac{\rho'}{\rho} &= \tfrac{1}{2}K'_z + K_z\frac{\rho'}{\rho} . \end{aligned} \tag{5.22}$$

The left-hand sides of these two equations are identical to the electrostatic equivalents of Eq. (5.4). We concentrated on the first of these equations in electrostatics because we are most interested there in meridional electrons (which have $\phi' = 0$ everywhere). The circular motion is much more important in magnetostatics and consequently we focus our attention here upon the second of Eqs. (5.22). Multiply it by ρ^2 to obtain

$$\frac{d}{dz}\left[\rho^2\left(\phi' - \frac{1}{2} K_z\right)\right] = 0 . \tag{5.23}$$

The mathematical solution of this equation is

$$\phi'(z) = \frac{1}{2} K_z + \frac{C}{\rho^2} , \tag{5.24}$$

where C is a constant of integration to be determined by the physical situation. One interesting application of (5.24) is to a case in which an electron is launched in an $\vec{H}_z$-field-free region, enters a region in which $H \neq 0$, and then exits into another field-free region. If the electron is initially meridional, then $\phi'(z) = 0$ so that $C = 0$, and the electron will appear to have been emitted from a point on the axis, provided $\rho' > 0$ initially. Of course C remains zero, according to (5.24), but K_z builds up to a nonzero value and the electron starts to rotate around the z axis as the electron enters into the region where $\vec{H}_z \neq 0$. Finally, $K_z = 0$ again when the electron exits this region and the electron is meridional again. It will intersect the z axis once again if $\rho'(z) < 0$. This example is sketched in Fig. 5-1 for the case in which an electron initially moves away from the

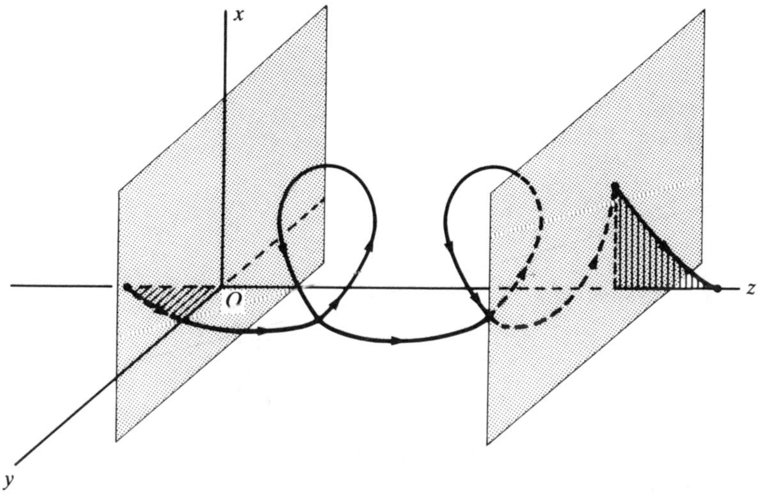

Fig. 5-1. An electron trajectory subject to Eq. (5.24) traversing a region of nonzero magnetic field.

axis in the $x = 0$ plane, then spirals around the axis as shown, and finally is meridional again in the $y = 0$ plane. The spiralling occurs in the region where $H_z \neq 0$.

5.4 MAGNETRON-LIKE DEVICES

The essential electron-optical feature of a magnetron is the existence of a *radial* electric field and a constant *axial* magnetic field (see Fig. 5-2). It is more useful here to analyze it directly in terms of the time as independent coordinate. In terms of nonrelativistic equations the accelerations are

$$\begin{aligned} \ddot{x} &= -\frac{e}{m}\,[E_x + (\dot{\vec{r}} \times \vec{B})_x]\,, \\ \ddot{y} &= -\frac{e}{m}\,[E_y + (\dot{\vec{r}} \times \vec{B})_y]\,, \end{aligned} \tag{5.25}$$

and these can be easily converted into the polar forms by manipulations similar to (5.3) to obtain

$$\begin{aligned} \ddot{\rho} - \rho\dot{\phi}^2 &= -\frac{e}{m}\,(E_\rho + \rho\dot{\phi}B_z - \dot{z}B_\phi)\,, \\ \frac{d}{dt}\,(\rho^2\dot{\phi}) &= -\frac{e}{m}\,\rho(E_\rho + \dot{z}B_\rho - \dot{\rho}B_z)\,. \end{aligned} \tag{5.26}$$

In the special magnetron case that $\vec{E} = E_\rho\hat{\rho}$ and $\vec{B} = B_z\hat{z}$, the second equation simplifies to

$$\frac{d}{dt}\,(\rho^2\dot{\phi}) - \frac{eB_z}{m}\,\rho\dot{\rho} = 0\,, \tag{5.27}$$

and, if B_z is also time-independent, to

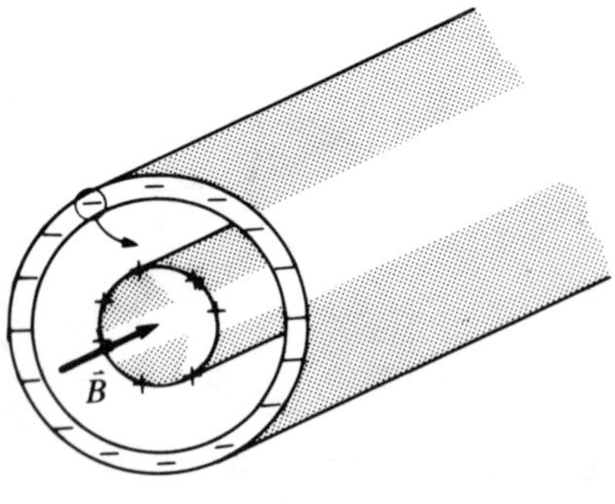

Fig. 5-2. A magnetron-like cylindrical capacitor emitting electrons in an axial magnetic field.

$$\frac{d}{dt}\left(\rho^2\dot{\phi} - \frac{1}{2}\,\Omega\rho^2\right) = 0\,, \tag{5.28}$$

where $\Omega \equiv eB_z/m$ is a rotation frequency (sometimes known as the cyclotron frequency) which expresses the rate of rotation of an electron around a constant magnetic flux density B_z. The radius of this rotation is given by $v_T/\Omega = K^{-1}$, if v_T is the magnitude of the velocity perpendicular to the axis. We first encountered K in Section 3.1, and we extended the concept there somewhat by defining $K_z = eB_z/mv_z$ even though the vector $\vec{v}_z$ changes direction only when there are nonzero B_x and/or B_y fields.

Electrons emitted by the outer sheath (the cathode) experience a range of thermal velocities that introduce small deviations from the radial direction into which they are pulled by the positive anode potential of the inner concentric conductor. If these small thermal velocities are ignored, then the initial $\dot{\phi}$ is zero at the cathode radius. If not, then the electron will rotate either towards or away from the anode. If away, then there will be some ρ_0 at which $\dot{\phi} = 0$. That ρ_0 serves as the reference radius for solving (5.28) which then leads to

$$\dot{\phi}(\rho) = \frac{1}{2}\,\Omega\left(1 - \frac{\rho^2}{\rho_0^2}\right)\,, \tag{5.29}$$

which enables us to obtain the angular velocity at the anode by setting $\rho = \rho_a$. We may approximate $\rho \approx \rho_c$ (cathode radius) for sufficiently small thermal velocities, in which case (5.29) yields an angular velocity $\dot{\phi}$ that can be calculated explicitly as a function of ρ. Further details on the ensuing electron trajectories in this device are given by Pierce (1949).

5.5 BUSCH'S THEOREM

Equation (5.29) is a special case of a theorem due to Busch and Brüche (1937) for the motion of electrons in a cylindrically symmetric set of fields for which $E_\phi = 0$ and $B_\phi = 0$. The second of Eqs. (5.26) then simplifies after multiplication by dt to

$$d(\rho^2\dot{\phi}) = \frac{e}{m}\,\rho(B_\rho\,dz - B_z\,d\rho)\,. \tag{5.30}$$

Consider an axially symetric surface, obtained by rotating a meridional trajectory $\rho(z)$, as sketched in Fig. 5-3, around the z axis, and consider

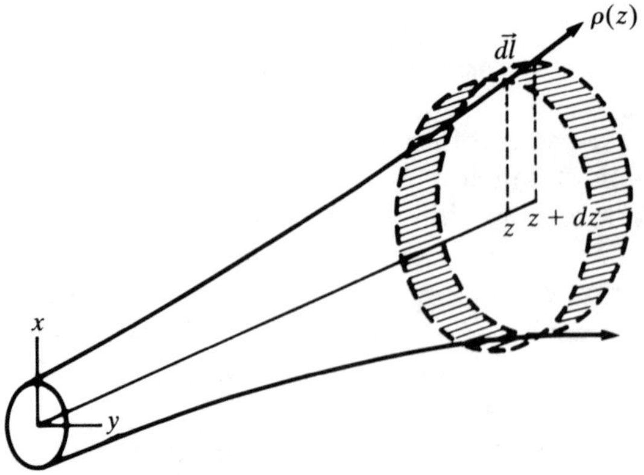

Fig. 5-3. Rotationally symmetric bundle of electron trajectories in an axisymmetric $\vec{B}$ field for discussion of Busch's theorem.

specifically the surface element $d\vec{S} = d\vec{l} \times \rho\hat{\phi}\, d\phi$, where $d\vec{l} = \hat{\rho}\, d\rho + \hat{z}\, dz$. The magnetic flux through the surface element is

$$\begin{aligned} d\Psi &= \vec{B} \cdot d\vec{S} = \rho\vec{B} \cdot (d\vec{l} \times \hat{\phi})\, d\phi = \rho(\vec{B} \times d\vec{l}) \cdot \hat{\phi}\, d\phi\,. \\ &= \rho(B_z\, d\rho - B_\rho\, dz)\, d\phi\,, \end{aligned} \tag{5.31}$$

for angular width $d\phi$ in the $\hat{\phi}$ direction. Now integrate over all azimuthal angles to obtain

$$d\Psi(z) = 2\pi\rho(z)(B_z\, d\rho - B_\rho\, dz)\,, \tag{5.32}$$

for the magnetic flux through the surface formed by rotating line element dl around the axis. Comparison of (5.32) with (5.30) then yields

$$d(\rho^2\dot{\phi}) = -\frac{e}{2\pi m}\, d\Psi\,. \tag{5.33}$$

We can integrate this expression from the location z_0 where $\dot{\phi} = 0$ to z, to obtain

$$\dot{\phi}(z) = -\frac{e}{2\pi m\rho^2(z)}\, [\Psi(z) - \Psi(z_0)]\,. \tag{5.34}$$

This is Busch's theorem, and it pertains for example to a meridional electron entering a region where $\vec{B}$ is nonzero (although it does require that $B_\phi = 0$ everywhere). In the special case that $B_\rho = 0$ and B_z is constant we have $d\Psi = 2\pi B_z\, d\rho = d(\pi\rho^2 B_z)$, which leads directly to Eq. (5.29).

REFERENCES

Busch, H. and E. Brüche (1937) *Beitrage zur Elektronenoptik*, Barth, Leipzig.

Goodman, J. W. (1986) *Introduction to Fourier Optics*, 2nd edn., McGraw-Hill, New York.

El-Kareh, A. B. and J. C. J. El-Kareh (1970) *Electron Beams, Lenses, and Optics*, Academic Press, Orlando, FL.

Pierce, J. R. (1949) *Theory and Design of Electron Beams*, Van Nostrand, Princeton, NJ.

CHAPTER 6

ELECTRON LENSES

The paraxial approximations discussed in Chapter 5 are sufficient for analyzing a large class of (idealized) electron lenses, especially if there is sufficient symmetry. Perhaps the most significant fact we shall learn is that cylindrically symmetric electrostatic or magnetostatic fields give rise to lensing (focussing or defocussing) actions. The same holds for planar symmetry. We will discuss general properties mainly of electrostatic lenses. Discussion of elementary electrostatic and magnetostatic lenses is given in Chapter 7.

6.1 ELECTROSTATIC LENSES—CYLINDRICAL SYMMETRY

The paraxial approximation to the equations of motion for $\rho(z)$ in an axisymmetric electrostatic field was given in Eq. (5.10), which is repeated here, as

$$\Phi_0\rho''(z) + \tfrac{1}{2}\Phi_0'\rho'(z) + \tfrac{1}{4}\Phi_0''\rho(z) = 0\,, \qquad (6.1)$$

in the understanding that $\Phi_0(z)$ is the *accelerating* potential (formerly denoted as $\Phi_{a0}(z)$) discussed in connection with Eqs. (2.26)–(2.29). This second-order total differential equation has two independent solutions $\rho_1(z)$ and $\rho_2(z)$ which are each determined by specific initial conditions. However, any two independent solutions obey an invariance (or conservation) property that we write as

$$\frac{d}{dz}\{\sqrt{\Phi_0(z)}[\rho_1'(z)\rho_2(z) - \rho_1(z)\rho_2'(z)]\} = 0\,. \tag{6.2}$$

This conservation or invariance of the $\{\ldots\}$ quantity in (6.2) is known in the theory of linear second-order differential equations as the *Sturm invariant*, and it is easily obtained from (6.1) by subtracting ρ_2 times the paraxial equation for ρ_1 from ρ_1 times the paraxial equation for ρ_2. Specifically, we obtain

$$\begin{aligned}\rho_2(\Phi_0\rho_1'' + \tfrac{1}{2}\Phi_0'\rho_1' + \tfrac{1}{4}\Phi_0''\rho_1) - \rho_1(\Phi_0\rho_2'' + \tfrac{1}{2}\Phi_0'\rho_2' + \tfrac{1}{4}\Phi_0''\rho_2)\\ = \Phi_0(\rho_1''\rho_2 - \rho_1\rho_2'') + \tfrac{1}{2}\Phi_0'(\rho_1'\rho_2 - \rho_1\rho_2') = 0\,.\end{aligned} \tag{6.3}$$

Divide the last line by $\sqrt{\Phi_0(z)}$ to obtain the equivalent of (6.2) after performing the differentiation explicitly.

The invariant (6.2) can be expressed in a physically more pleasing way by using the momentum representation of Section 3.2. The transverse momentum of Eq. (3.27) can be written for axisymmetric potentials in the paraxial approximation as

$$p_T \approx \sqrt{2em\Phi_0}\,\rho'\,. \tag{6.4}$$

It is advantageous, once again, to define $P_T = p_T/\sqrt{2m_0}$; this new quantity is proportional to momentum but it has the dimension of $\sqrt{\text{energy}}$ and is often measured in $\sqrt{\text{eV}}$. We followed a similar convention in connection with the trajectory equations (3.60). The paraxial equation then gives us, instead of (6.4), the more compact form,

$$P_T \approx \sqrt{e\Phi_0}\,\rho'\,, \tag{6.5}$$

which makes quite explicit that P_T is measured in units of $\sqrt{\text{eV}}$ when Φ_0 is expressed in volts. Finally, if we multiply (6.2) by the electron charge e and then substitute (6.5) into it, we obtain the determinant equation,

$$\frac{d}{dz}\left\{\det\begin{bmatrix}\rho_1(z) & \rho_2(z)\\ P_{T1}(z) & P_{T2}(z)\end{bmatrix}\right\} = 0\,. \tag{6.6}$$

Expressions (6.2) and (6.6) are equivalent; sometimes one and sometimes the other is handier in use.

Let us now use the above equations to understand imaging in an axisymmetric electrostatic field given by potential $\Phi(\rho, z)$. Consider the two special solutions $\rho_1(z)$ and $\rho_2(z)$ of the paraxial equation sketched in Fig. 6-1. The electron with trajectory $\rho_1(z)$ is launched parallel to the z

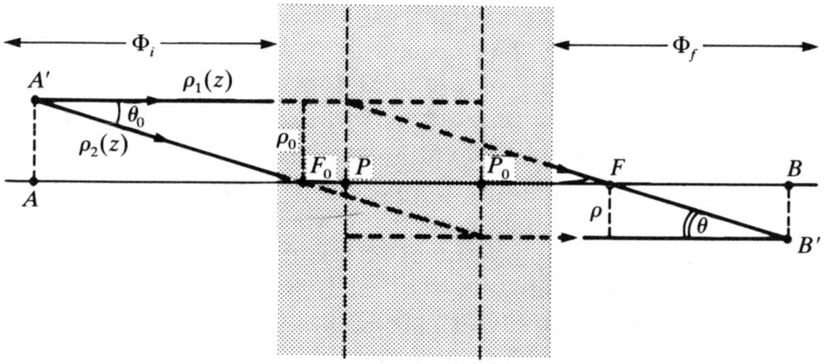

Fig. 6-1. The classical construction of principal planes and focal points for a thin lens in the paraxial-optics approximation.

axis at $z = -\infty$, and it passes through point A' at ray height ρ_0. The potential is a constant value Φ_0 at least up to point A'. The electron then enters the lensing region in which a transition of potential from an initial value Φ_i to a final value Φ_f is made. Beyond the lensing region, where the potential is again at a constant value, Φ_f, the electron maintains a slope with ray-coordinate derivative $\rho'(z) = -\tan\theta$, and it then passes through point B' with this slope. The *image focal point D* is then obtained by extrapolating the ray coordinate $\rho(z)$ back (at constant slope) until the z axis is intersected. The assumption is made that point F lies outside the lensing region. Likewise, the *image principal plane* is defined by first finding the point P by intersecting the extension of the line $B'F$ with the horizontal ray $\rho(z) = \rho_0$. The image principal plane is perpendicular to the z axis and it goes through point P.

The corresponding *object* focal point and principal plane are found by tracing a second electron with trajectory $\rho_2(z)$ and slope $\rho_2'(z) = 0$ *backwards* from $z = \infty$. On the object side of the lensing region, the slope will be $\rho_2(z) = -\tan\theta_0$ for all z to the left of point A (and possibly for some z to the right of A). The intersection with the z axis gives the *object focal point* F_0 which also is assumed not to lie in the lensing region. Incidentally, the ray height $\rho_2(z) = -\rho$ has been chosen so that the second electron also passes through point A'. The point B' is then determined by intersecting the two electron trajectories to the right of the lensing region. Finally, the *object principal plane* is determined by point P_0 which is the intersection of the extension of $\rho_2(z)$ for $z < z_A$ with $\rho_2(z) = -\rho$.

The situation in which either or both of the focal points F_0, F lie inside the lensing region pertains to so-called *immersion lenses*. These are more difficult to analyze. We include the *thin lens* among those cases for which F_0 and F both lie outside the lensing region, but a thin lens requires more stringent conditions for the approximations valid for it.

Let us first note that the ray heights ρ_0 and ρ_1 are arbitrary in defining the four points F_0, P_0, F, P. Choose, for example, two new trajectories $\tilde{\rho}_1(z) = \alpha_1 \rho_1(z)$ and $\tilde{\rho}_2(z) = \alpha_2 \rho_2(z)$. At focal points F_0 and F both new trajectory coordinates are proportional to the old ones, which are zero. Hence the new trajectories also cross the z axis at F_0 and F. The fact that the ray height $\rho(z) = \rho_0$ and the ray tangent $\rho_1(z) = -\tan\theta$ increase by factor α_1, and that point F is fixed, shows that point P is also fixed. Likewise we show that P_0 is fixed. The invariance of points F_0, P_0, F, P is, of course, a consequence of the linearity of the paraxial equations. These four important points are therefore entirely determined by the profile of the potential $\Phi(z)$ and they are independent of whatever trajectories we choose to obtain them. Their role will be explained further as we proceed.

We shall now derive the socalled *thin-lens formulas*. Apply (6.2) to the region between F_0 and F to obtain

$$\mathbb{S}(F) = \mathbb{S}(F_0) \quad \text{with} \quad \mathbb{S}(z) \equiv \sqrt{\Phi_0(z)}[\rho_1'(z)\rho_2(z) - \rho_1(z)\rho_2'(z)] \,. \tag{6.7}$$

When the values of ray height and slope are inserted at F_0 and F, we obtain from (6.7),

$$\sqrt{\Phi_i}\rho_0 \tan\theta_0 = \sqrt{\Phi_f}\rho \tan\theta \,, \tag{6.8}$$

which is a paraxial form of what is known in optics as *Abbe's sine law* (Born and Wolf 1965). The sines of the angles rather than the tangents usually appear in Abbe's sine law, but of course $\tan\theta \approx \sin\theta \approx \theta$ in the paraxial approximation so that the difference is inconsequential here. Referring again to the sketch in Fig. 6-1, we define the following distances:

$f_0 = F_0P$, object focal distance,

$f = FP_0$, image focal distance,

$L_0 = AP_0$, object distance,

$L = BP$, image distance.

The obvious geometrical relationship $\rho/\rho_0 = FB/FP = F_0P_0/F_0A$ yields the *magnification*,

$$\frac{\rho}{\rho_0} = \frac{L-f}{f} = \frac{f_0}{L_0 - f_0} \,. \tag{6.9}$$

We obtain $(L_0 - f_0)(L - f) = f_0 f$ from this, and after division by $L_0 L$ we obtain the *thin-lens* formula that relates image to object distance,

$$\frac{f_0}{L_0} + \frac{f}{L} = 1 \, . \tag{6.10}$$

We should bear in mind carefully that the distances f_0, f, L_0, L have all been defined as *positive* quantities. Some authors of texts and review articles in optics deviate from this convention, but we find it more convenient to work with positive lengths.

Another important paraxial law is obtained from the Abbe sine law as given by (6.8) after replacing tan θ_0 by ρ/f_0 and tan θ by ρ_0/f (as indicated by the geometry in Fig. 6-1), and after dividing out a common factor $\rho_0\rho$,

$$\frac{f}{f_0} = \sqrt{\frac{\Phi_f}{\Phi_i}} \, . \tag{6.11}$$

This relationship between focal distances and field-free potentials is known as the *Lagrange-Helmholtz* relationship. If it is substituted into the thin-lens formula (6.10), we obtain the important formula,

$$\frac{\sqrt{\Phi_i}}{L_0} + \frac{\sqrt{\Phi_f}}{L} = \frac{\sqrt{\Phi_i}}{f_0} = \frac{\sqrt{\Phi_f}}{f} \equiv s \, . \tag{6.12}$$

In Chapter 2 we observed that $\sqrt{\Phi}$ is the electron optics equivalent of a refractive index. Equation (6.12) therefore has exactly the same form as the analogous thin-lens formula for light optics lenses made of glass or other transparent refractive index media. The ratio $\sqrt{\Phi_i}/f_0 = \sqrt{\Phi_f}/f \equiv s$ is known as the *strength* of the lens, and it is measured in $\sqrt{\text{V}}$/m. We shall see that it is closely related to the transverse momentum of an electron that has been accelerated from zero or negligible energy.

So far, we have worked out explicitly a number of thin-lens formulas for imaging. We have seen that a complete analogy with light optics holds, and consequently we could take the point of view that all of the theorems of geometrical optics hold and simply refer the reader to optics texts. However, we do wish to make this text as self-contained as possible, and therefore we will work out independently some of the more important consequences for electron optics.

Referring as an example to Fig. 6-1, we will now show that point B' is the *image* of point A'; i.e., that any electron trajectory moving through point A' towards the lens will also go through point B', no matter what its slope is at A'. We again shall need the special trajectories $\rho_1(z)$ and $\rho_2(z)$

that we defined previously. Define also $\delta\rho(z) \equiv \rho_1(z) - \rho_2(z)$. Quite obviously, $\delta\rho(z)$ passes through the points A and B because $\rho_1(z_A) = \rho_2(z_B) = 0$. Finally define α as an arbitrary real constant and define a new trajectory,

$$\rho_3(z) \equiv \rho_1(z) + \alpha\delta\rho(z) = (1+\alpha)\rho_1(z) - \alpha\rho_2(z) . \tag{6.13}$$

If we look at the value of this new electron trajectory coordinate and its derivative at points A and B, we obtain

$$\begin{aligned} \rho_3(z_A) &= (1+\alpha)\rho_1(z_A) - \alpha\rho_2(z_A) = \rho_0 , \\ \rho_3'(z_A) &= (1+\alpha)\rho_1'(z_A) - \alpha\rho_2'(z_A) = \alpha \tan\theta_0 , \\ \rho_3(z_B) &= (1+\alpha)\rho_1(z_B) - \alpha\rho_2(z_B) = -\rho . \end{aligned} \tag{6.14}$$

This states that the trajectory $\rho_3(z)$ passes through point A' with derivative or slope given by $\tan\theta_3 \equiv \alpha \tan\theta_0$, and it proves that the trajectory also passes through point B'. Because $\rho_3(z)$ is an arbitrary trajectory through A', this proves that B' is the image of A'.

The general shape of a trajectory such as $\rho_1(z)$ in a lensing region in which $\Phi(z)$ increases smoothly and monotonically from Φ_i to Φ_f is sketched in Fig. 6-2. The transition from Φ_i to Φ_f gives rise to equipotential surfaces Φ_1 and Φ_2 with $\Phi_i < \Phi_1 < \Phi_2 < \Phi_f$. The sketched shapes of these equipotential surfaces follows from the paraxial approximation (5.6) together with the polarity of $\Phi_0''(z)$ (which is initially positive and becomes negative near the exit of the lensing region. A *principal electron* (i.e., an electron with initial $\rho'(z) = 0$) will accelerate at potential Φ_1 and the concomitant electrostatic field $\vec{E}_1$ will deflect it towards the axis. As the electron passes through the Φ_2 equipotential surface it undergoes an $\vec{E}_2$ field that deflects it away from the axis. Closer scrutiny, e.g., with Eqs.

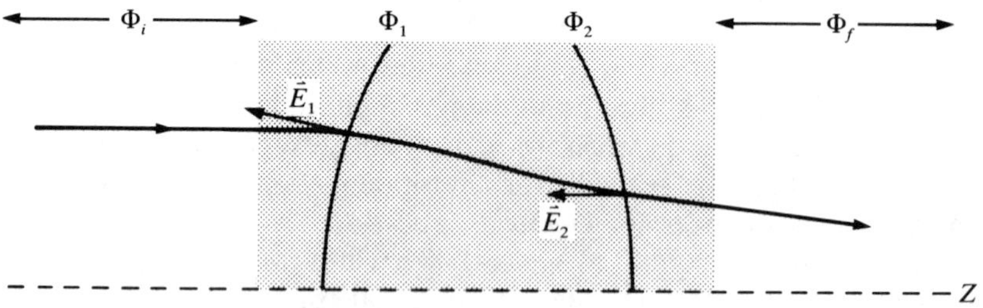

Fig. 6-2. An electron trajectory in an accelerating-lens region. Notice focussing if initial and final regions are field-free.

(5.8), will reveal that the transverse field E_{2T} is less than E_{1T}, and, moreover, the electron is moving faster at Φ_2 in the z direction. These two reasons combine to prevent the deflection away from the axis from being as effective as the initial deflection towards the axis; the net result is a deflection *towards* the axis.

A number of textbooks on electron optics also show sketches of trajectories that first deflect away from the axis but end up with a total deflection towards the axis. The reason is that the $\Phi_0(z)$ profile is more complicated than or different from the one above. One example is the *unipotential* (or *einzel*, in the pre-war German literature) lens, in which $\Phi_f = \Phi_i$, but there is an intermediate lower potential value Φ_{im} (created, for example, by an intermediate cylinder held at voltage Φ_{im}).

6.1.1 Solution of the Paraxial Equation

The treatment of *thick* lenses (e.g., immersion lenses) is more difficult, and it would be necessary to solve the paraxial equation to do so properly. It is rarely possible to find an analytical solution, but (6.1) can be cast into a generalization of (6.12). To do so, we divide each term of (6.1) by $\sqrt{\Phi_0(z)}$ to obtain

$$\sqrt{\Phi_0}\rho''(z) + \frac{1}{2}\frac{\Phi_0'}{\sqrt{\Phi_0}}\rho'(z) = -\frac{1}{4}\frac{\Phi_0''}{\sqrt{\Phi_0}}\rho(z)\,,$$

or

$$\frac{d}{dz}(\sqrt{\Phi_0}\rho') = -\frac{1}{4}\frac{\Phi_0''}{\sqrt{\Phi_0}}\rho(z)\,, \tag{6.15a}$$

and integration from $z = 0$ to an arbitrary value of $z > 0$ yields

$$\sqrt{\Phi_0(z)}\rho'(z) - \sqrt{\Phi_0(0)}\rho'(0) = -\frac{1}{4}\int_0^z d\zeta\,\frac{\Phi_0''(\zeta)}{\sqrt{[\Phi_0(\zeta)]}}\,\rho(z)\,. \tag{6.15b}$$

We can write this in an abbreviated form, if 0 and z lie in field-free end regions, as

$$\sqrt{\Phi_f}\rho' - \sqrt{\Phi_i}\rho_0' = -\frac{1}{4}\int_0^z d\zeta\,\frac{\Phi_0''(\zeta)}{\sqrt{[\Phi_0(\zeta)]}}\,\rho(z)\,. \tag{6.15c}$$

The first question that arises is that of how we might construct the image point B of a trajectory starting at some point A on the z axis in a field-free region. If the lens is sufficiently weak (so that deflection of a trajectory is slight) then we can be reasonably sure that point B is also in a field-free region: "zero" weakness would mean no lensing action whatever and point B would be at $z = \infty$ if anywhere. The right-hand side of (6.15c) is associated with the strength of the lens. Under the assumption again that 0 and z are in field-free end regions, we define $\rho_0' = \tan\theta_0$ and $\rho' = -\tan\theta$. Note the plus sign in front of $\tan\theta_0$ because we are now trying to image a point on the z axis, which implies that we aim the electron away from the axis, i.e., up, initially. Let us now define $\tan\theta_0 \equiv \rho_L/L_0$ and $\tan\theta \equiv \rho_L/L$ where they ray height ρ_L, and the two distances L_0, L will be specified below. This transforms (6.15b) into

$$\frac{\sqrt{\Phi_f}}{L} + \frac{\sqrt{\Phi_i}}{L_0} = \frac{1}{4}\int_0^z d\zeta \, \frac{\Phi_0''(\zeta)}{\sqrt{[\Phi_0(\zeta)]}} \, \frac{\rho(\zeta)}{\rho_L} \, . \tag{6.16}$$

Comparison with the thin-lens formula (6.12) shows that the expression on the right-hand side of (6.16) is the strength factor s, which is independent of $\rho(z)$ in the thin-lens case. Thus, the necessary condition for a lens to be considered *thin* is that $\rho(\zeta) \approx \rho_L$ throughout the entire lensing region. A lens is not necessarily thin if points A and B lie outside the lensing region.

When all of the factors discussed above are taken into account, then the construction of the rays through the lensing region is as shown in Fig. 6-3. The only new feature in electron optics is that the principal planes P_0

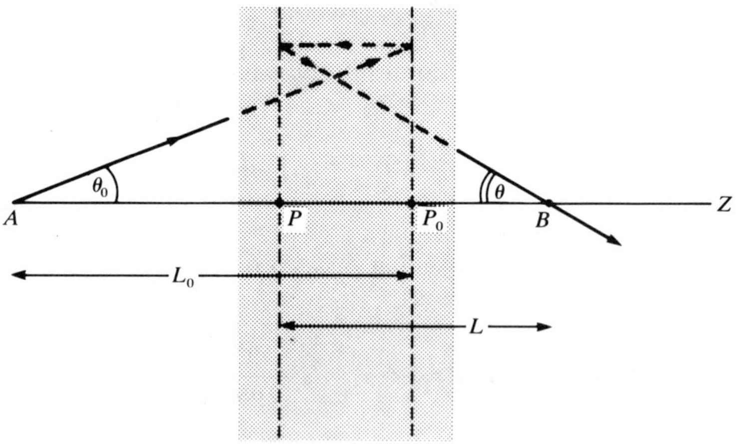

Fig. 6-3. Imaging through a thin lens in the paraxial-optics approximation.

and P are *reversed* (compared to light optics); the image principal plane precedes the object one! We shall discuss this carefully later, but let us first stress that the *actual* trajectory connecting points A and B has not been sketched in Fig. 6-3. The sketch is correct only in the field-free regions outside the lensing region. Furthermore, the lensing region does not necessarily coincide with the region between the principal planes P_0 and P; in fact either P_0 or P can lie well into a field-free region. Nevertheless, once points P_0 and P have been constructed, as discussed in connection with Fig. 6-1 by means of the two special principal-electron trajectories, then an image point B can be constructed as shown from given values of $\rho(0)$ and $\rho'(0)$ at a point A in a field-free region.

If the lens is thick, and its strength (given by the right-hand side of (6.16)) is sufficient to immerse either B or A in the lensing region, then the above construction will fail. We repeat, both A and B need to be in field-free regions.

For the case of a thin lens, with $\rho(\zeta) \approx \rho_L$ in the lensing region, we find that (6.16) simplifies to

$$\frac{\sqrt{\Phi_f}}{L} + \frac{\sqrt{\Phi_i}}{L_0} = \frac{1}{4} \int_0^z d\zeta \, \frac{\Phi_0''(z)}{\sqrt{[\Phi_0(\zeta)]}} \, . \tag{6.17}$$

Comparison with (6.12) tells us that the lens strength s is given explicitly by

$$s = \frac{1}{4} \int_{-\infty}^{\infty} d\zeta \, \frac{\Phi_0''(\zeta)}{\sqrt{[\Phi_0(\zeta)]}} \, . \tag{6.18}$$

Note that the extension of the bounds to $\pm\infty$ does not change matters, as derivatives of Φ_0 vanish outside the (entirely contained) lensing region. Further comparison of the two thin-lens formulas (6.12) and (6.17) allows us to write down explicit expressions for the focal lengths,

$$\begin{aligned} \frac{1}{f} &= \frac{1}{4\sqrt{\Phi_f}} \int_{-\infty}^{\infty} d\zeta \, \frac{\Phi_0''(\zeta)}{\sqrt{[\Phi_0(\zeta)]}} \, , \\ \frac{1}{f_0} &= \frac{1}{4\sqrt{\Phi_f}} \int_{-\infty}^{\infty} d\zeta \, \frac{\Phi_0''(\zeta)}{\sqrt{[\Phi_0(\zeta)]}} \, . \end{aligned} \tag{6.19}$$

6.1.2 Solution of the Picht Paraxial Equation

We have seen in Chapter 5 that the paraxial equation could be cast into an interesting alternative form in terms of the Picht-reduced radial coordinate $R(z) \equiv [\Phi_0(z)]^{1/4} \rho(z)$. We still retain the notation $\Phi_0(z)$ for the on-axis accelerating potential. Returning to Eq. (5.11), we integrate it from 0 to z to obtain

$$R'(z) - R'(0) = -\frac{3}{16}\int_0^z d\zeta\left[\frac{\Phi_0'(\zeta)}{\Phi_0(\zeta)}\right]^2 R(\zeta)\,. \tag{6.20}$$

How do the solutions (6.20) and (6.15) compare? One will sometimes find arguments in favor of (6.20) for certain common lenses, but it is not possible to favor one over the other in general. For *bipotential* lenses, defined by initial Φ_i and final Φ_f such that $\Phi_0(z)$ rises monotonically in between (see Fig. 6-4), a case can be made that the reduced coordinate $R(z)$ is more constant than $\rho(z)$ in the lensing region. The obvious reason is that the decrease in $\rho(z)$ (at least for principal-electron trajectories) is somewhat offset by a concomitant increase in the factor $[\Phi_0(z)]^{1/4}$ as z increases. Thus, for thick but weak lenses, it would be more likely that $R(\zeta)$ can be replaced by a constant R_L in (6.20), at least for many typical beam trajectories.

Application of (6.16) and (6.20) to a principal electron with $\rho_0' = 0$ and $\rho_f' = -\rho_L/f$ yields two formulas for the focal distance f:

$$\begin{aligned} \frac{1}{f} &= \frac{1}{4\sqrt{\Phi_f}}\int_{-\infty}^{\infty} d\zeta\,\frac{\Phi_0''(\zeta)}{\sqrt{[\Phi_0(\zeta)]}}\,\frac{\rho(\zeta)}{\rho_L}\,, \\ \frac{1}{f} &= \frac{3}{16}\int_{-\infty}^{\infty} d\zeta\left[\frac{\Phi_0'(\zeta)}{\Phi_0(\zeta)}\right]^2 \frac{R(z)}{R_L}\,. \end{aligned} \tag{6.21}$$

The first of these formulas is numerically somewhat harder to integrate accurately, (i) because the second derivative $\Phi_0''(\zeta)$ is often less accurate

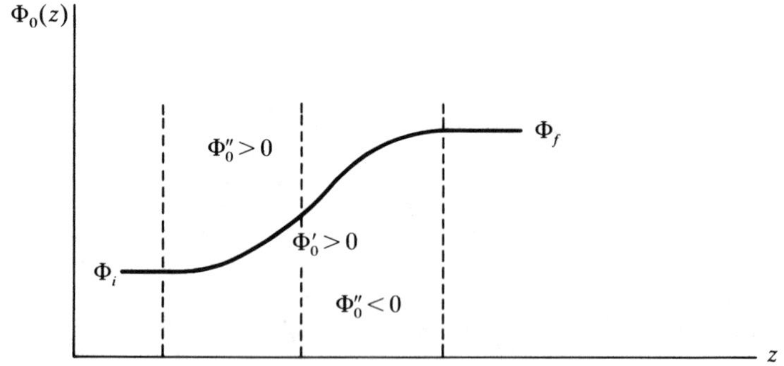

Fig. 6-4. The axial profile of the potential for a *bipotential* lens.

when it is obtained numerically, and (ii) because the second derivative changes sign. The second formula has the numerical advantage of having a positive-definite potential factor in the integrand. Nevertheless, the ultimate decision in using one or the other of these formulas may rest on whether $\rho(\zeta)$ or $R(\zeta)$ is closer to being constant in the lensing region. Grivet (1972) discusses other reasons for preferring the Picht formula, but we stress that it is not clear a priori which is to be preferred.

6.1.3 Accelerating Lenses Always Focus

We shall now prove that axisymmetric lensing regions lead to focussing when $\Phi_f \geqslant \Phi_i$. Consider an electron trajectory that does not cross the axis in the lensing region and let $\rho(z)$ be positive there. According to the Picht formula (6.20), we then have

$$R'_f - R'_0 = -\frac{3}{16}\int_{-\infty}^{\infty} d\zeta \left[\frac{\Phi'_0(\zeta)}{\Phi_0(\zeta)}\right]^2 R(\zeta)\,, \tag{6.22}$$

if initial and final points lie in field-free regions. The integrand is obviously positive because $R(\zeta)$ is, and consequently we deduce that $R'_f < R'_0$, which in turn implies that

$$\rho'_f < \left(\frac{\Phi_i}{\Phi_f}\right)^{1/4} \rho'_0\,. \tag{6.23}$$

It must follow that $\rho'_f < \rho'_0$ because $\Phi_i \leqslant \Phi_f$, and therefore that the electron is deflected *towards* the axis. On the other hand, if $\Phi_i \gg \Phi_f$, then the electron is decelerated and it is deflected away from the axis. If its initial kinetic energy in insufficient to overcome the barrier of low Φ_f then it will actually deflect by more than 90° and thus turn back towards its region of origin. Low potentials can behave as *electron mirrors*.

A useful analogy to the behavior of electrons in an electrostatic field is the now seldom-used rubber-sheet model. Imagine the electron to be a small steel ball placed (possibly with initial motion) on a large rubber sheet. Negative potential (which decreases the electron's velocity) corresponds to hills, and positive potential corresponds to dips or valleys in the sheet. Note that "positive" or "negative" is relative to the potential defined by the initial kinetic energy measured in electron volts (eV). The trajectory of the steel ball will then ideally be the same as that of an electron in a field where the potential at every point is proportional to the relative height of the rubber sheet compared to a location where the ball was initially at rest. The rubber sheet was a useful, although gross, tool before the advent of the digital electronic computer, but the *mental* comparison is still useful for the student of electron optics. For example, by conjuring up the mental image of a rubber sheet with dips and peaks, one can easily picture the focussing effect described above in connection with (6.22) and (6.23).

Focussing can still occur when $\Phi_f \leqslant \Phi_i$. This is contained in (6.23) as that inequality predicts $\rho_f' < \rho_0'$ when $\Phi_f = \Phi_i$ (as in unipotential lenses). As one considers situations with Φ_f increasingly smaller than Φ_i, we reach a situation where $\rho_f' = \rho_i'$ (assuming that we do have a lensing region; otherwise $\Phi_f = \Phi_i$ would give $\rho_f' = \rho_0'$ trivially). Hence, as long as the right-hand side of (6.22) is nonzero we will reach the situation at which $\rho_f' = \rho_0'$ at some $\Phi_f \leqslant \Phi_i$. For even lower Φ_f we have defocussing.

6.1.4 The Principal Plane are Always Crossed

Unlike ordinary light optics lenses, electrostatic lenses have principal planes that are always crossed. This is most easily understood by reference to the Picht formula (6.22) and the reduced trajectory coordinate $R(z)$. We may assume, for a thin lens, that $R(\zeta)$ is essentially constant in the lensing region for trajectories that enter or exit with $R'(z) = 0$. Consider Fig. 6-5 and the three trajectories sketched in it (all are in terms of the reduced Picht radial coordinate $R(z)$). The first, $R_1(z)$, emanates from A' or A_1' and goes through points B' and B_1'. The second, $R_2(z)$, emanates from A' and goes through B' with zero derivative. The third, $R_3(z)$, is run backwards with zero derivative from B_1' such that it goes through A_1'. Because the Picht solution for a thin lens is

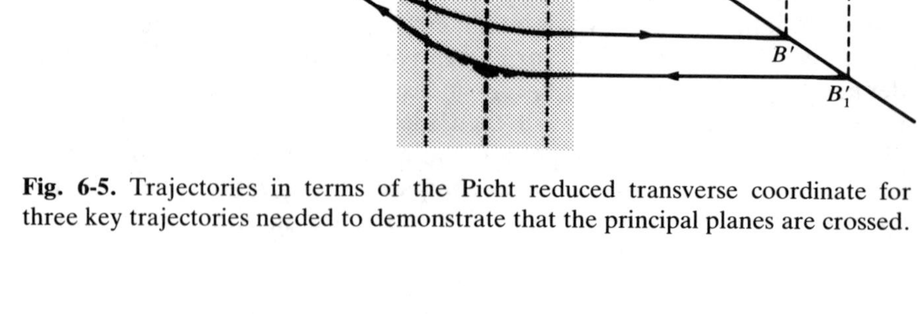

Fig. 6-5. Trajectories in terms of the Picht reduced transverse coordinate for three key trajectories needed to demonstrate that the principal planes are crossed.

$$R_f' - R_0' \approx -R_L S ,$$

$$S \equiv \frac{3}{16} \int_{-\infty}^{\infty} d\zeta \left[\frac{\Phi_0'(\zeta)}{\Phi_0(\zeta)} \right]^2 ,$$

it follows that trajectories #1 and #3 are mirror images of each other (if one is mirrored first across the z axis and then across a plane, perpendicular to the z axis, through $\bar{P}$). In fact, this helps define point $\bar{P}$, and the two dashed vertical lines represent planes that essentially terminate the lensing region. Trajectory #2 obeys

$$R_2(z) = \alpha R_3(z) , \qquad R_2'(z) = \alpha R_3(z) , \tag{6.25}$$

everywhere where $\alpha < 1$ is a constant. Equations (6.25) and the above-described mirroring property define $\bar{P}$ as the intersection of a common principal plane for $R(z)$ trajectories. We next define $\bar{f}_0 = \bar{P}\bar{F}_0$ and $\bar{f} = \bar{P}\bar{F}$ as object and image focal distances in terms of the $R(z)$ coordinate. The remainder of the proof that the $\rho(z)$ principal planes are crossed follows the argument given by El-Kareh and El-Kareh (1970), and it goes as follows:

Apply (6.24) to $R_1(z)$ and $R_2(z)$ to obtain

$$R_1(\infty) = -R_1(-\infty)S , \qquad R_2(-\infty) = R_1(\infty)S , \tag{6.26}$$

from which we observe that $1/S = \bar{f} = \bar{f}_0$, with $\bar{f}_0$ to the left and $\bar{f}$ to the right of $\bar{P}$. We have made use of the fact that $R_1(\infty) > 0$ and that the other three R and R' quantities in (6.26) are negative to obtain positive focal distances. Note that S is always positive-definite.

Now make the translation into $\rho(z)$ quantities:

$$\begin{aligned} \bar{f} &= \frac{1}{S} = -\frac{R_1(-\infty)}{R_1'(\infty)} = -\left(\frac{\Phi_i}{\Phi_f}\right)^{1/4} \frac{\rho_1(-\infty)}{\rho_1'(\infty)} = \left(\frac{\Phi_i}{\Phi_f}\right)^{1/4} f , \\ \bar{f}_0 &= \frac{1}{S} = -\frac{R_1(\infty)}{R_2'(-\infty)} = -\left(\frac{\Phi_f}{\Phi_i}\right)^{1/4} \frac{\rho_2(\infty)}{\rho_2'(-\infty)} = \left(\frac{\Phi_f}{\Phi_i}\right)^{1/4} f_0 . \end{aligned} \tag{6.27}$$

From this, we observe for accelerating lenses ($\Phi_f \geqslant \Phi_i$) that $f_0 \leqslant \bar{f}_0$ and $f \geqslant \bar{f}$ with $\bar{f}_0 = \bar{f}$. Because the focal points F_0 and F are the same for $\rho(z)$ and for $R(z)$ trajectories, it follows that length $F_0P_0 = f_0$ is smaller than length $\bar{F}_0\bar{P} = F_0\bar{P} = \bar{f}_0$, hence point P_0 lies to the left of $\bar{P}$. Likewise $FP = f$ is larger than $\bar{F}\bar{P} = F\bar{P} = \bar{f}$ so that point P also lies to the left of $\bar{P}$. Finally, use

$$\bar{P}P = FP - F\bar{P} = f - \bar{f} = \left[\left(\frac{\Phi_f}{\Phi_i}\right)^{1/4} - 1\right]\bar{f}\,,$$
$$\bar{P}P_0 = F_0\bar{P} - F_0\bar{P} = \bar{f}_0 - f_0 = \left[1 - \left(\frac{\Phi_i}{\Phi_f}\right)^{1/4}\right]\bar{f}_0\,. \tag{6.28}$$

It is obvious that $\bar{P}P > \bar{P}P_0$ when $\Phi_f > \Phi_i$, hence for *accelerating* lenses P lies on the low-potential side of P_0 which proves that the principal planes are then crossed.

6.2 MAGNETOSTATIC LENSES—CYLINDRICAL SYMMETRY

We will only treat magnetostatic lenses cursorily, as the axisymmetric ones are governed by equations similar to those for electrostatic lenses. We have already discussed magnetron-like devices with axial fields in Section 5.4. We found magnetostatic paraxial equations, which we expressed in polar coordinates in Eqs. (5.22) and (5.34), and which are repeated here for convenience:

$$\rho'' - \rho(\phi')^2 = -K_z\rho\phi'\,,$$
$$\frac{d}{dz}\,[\rho^2(\phi' - \tfrac{1}{2}K_z] = 0\,, \tag{6.29}$$

where $K_z = eB_z/mv_z$. If the magnetic flux density has no transverse component ($B_T = 0$), and if the electric field is purely transverse ($E_z = 0$), then the axial velocity v_z will be a constant of the motion, and K_z varies only if $B_z(\rho, z)$ varies with ρ and/or z. The solution of the second of Eqs. (6.29) was given in Eq. (5.24). We showed that the special solution $\phi'(z) = \frac{1}{2}K_z(z)$ corresponds to trajectories of electrons that are initiated in one meridional plane, then rotate around the z axis with the rotation frequency $\phi' = \frac{1}{2}K_z$ after entering a $\vec{B}$-field region, and finally exit the $\vec{B}$-field region in another meridional plane. Now consider the first of Eqs. (6.29) after substitution of $\phi' = \frac{1}{2}K_z$ to obtain

$$\rho'' = -\tfrac{1}{4}K_z^2\rho\,. \tag{6.30}$$

This equation is analogous to the Picht equation (5.11) for the reduced radial coordinate, and therefore we obtain an expression for $\rho'(z)$ (see Eq. (6.32)) similar to (6.22). The major difference in the kinematics of the situation is that there is a simultaneous rotation around the z axis because $\phi'(z)$ is nonzero. To summarize this, we conclude that there are *two* independent solutions to (6.30), and the Sturm invariant (formed in the same way as (6.2)) is

$$\frac{d}{dz}\,[\rho_1'(z)\rho_2(z) - \rho_1(z)\rho_2'(z)] = 0\,. \tag{6.31}$$

Integration of Eq. (6.29) yields

$$\rho'(z) - \rho'(0) = -\frac{1}{4}\int_0^z d\zeta\; K_z^2(\zeta)\rho(\zeta)\,. \tag{6.32}$$

Hence the strength factor is the right-hand side of this equation, divided by ρ_L, which, for thin lenses, becomes

$$s = -\frac{1}{4}\int_0^z d\zeta\; K_z^2(\zeta)\,. \tag{6.33}$$

The entire thin-lens development of Eqs. (6.20)–(6.27) is also valid for the magnetostatic situation described by Eq. (6.32), and the fact that (6.30) yields the inequality $\rho''(z) \leqslant 0$ expresses convergence of this type of lens, i.e., "concave" trajectories. The analogy with electrostatic lenses is quite strong with one other major difference: the focussing strength s depends upon the axial electron velocity v_z, a fact which can give rise to large *chromatic aberrations* (errors in focal or image point because different electrons travel at different velocities). In light optics, chromatic aberrations are due to wavelength dependence of the refractive index which causes photons of different wavelength to deflect at different angles.

REFERENCES

Born, M. and E. Wolf (1965) *Principle of Optics*, 3rd edn., Pergamon, Oxford.

El-Kareh, A. B. and J. C. J. El-Kareh (1970) *Electron Beams, Lenses, and Optics*, Academic Press, Orlando, FL.

Grivet, P. (1972) *Electron Optics*, 2nd edn., Pergamon, Oxford.

CHAPTER 7

CLASSES OF ELECTRON LENSES

We discuss a number of practical, chiefly electrostatic, lenses that are used in cathode-ray tubes, electron microscopes, etc., in this chapter. Our interest will focus on the electron-optical properties of these lenses rather than on other matters such as method of construction. We shall also idealize the lenses by ignoring details such as the small (but nonzero) thickness of metal plates holding the defining voltages of the lens. The present classification is not exhaustive by any means; rather, it is our intention to show the reader how the principles explained up to now can be used to aid in the design of actual lenses, and also to show what can be done analytically before one feels forced to adopt an entirely numerical approach. The examples in this chapter show that analytical skills aid considerably in the analysis of electron optics lenses.

7.1 ELECTROSTATIC CYLINDER LENSES

The first class of electron lenses we shall consider consists of a series of N finite metallic cylinders around one axis (the z axis), each cylinder having the radius R. These cylinders are placed end-to-end with relatively small gaps between them. Such a configuration of cylinders, all with radius R, can model an axial potential distribution by appropriate choice of cylinder voltages V_i $(i = 1, \ldots, N)$. For purposes of analysis, we shall define $V(z) = \Phi(R, \phi, z)$, as the azimuthal coordinate ϕ plays no role, and the radial coordinate $\rho = R$ is fixed (see Fig. 7-1). Note that the interior region $\rho < R$ is not entirely enclosed, and that we do not really know the potential at $\rho = R$ in the gaps. If, however, the gap is small (compared to radius R), then it is reasonable to assume that $V(z)$ changes linearly between cylinders; there is little reason at any rate to expect that some deviations from a linear profile will have a significant effect when the gaps are small. The symmetry tells us that Laplace's equation becomes

$$\frac{1}{\rho}\frac{\partial}{\partial \rho}\left(\rho \frac{\partial \Phi}{\partial \rho}\right) + \frac{\partial^2 \Phi}{\partial z^2} = 0 . \qquad (7.1)$$

Please note that this is an *exact*, not a paraxial equation! It can be solved in principle by the separation-of-variables method (see Section 4.1.2). If the separation constant for the z part is $-k^2$, then a trial solution of (7.10) would be

$$\Phi(\rho, z) = I_0(k\rho)\exp(-ikz) , \qquad (7.2)$$

where $I_0(\ldots)$ represents the modified Bessel function of order 0 that is finite at $\rho = 0$ (Abramowitz and Stegun 1964). However, as we explained in Section 4.1.2, a solution such as (7.2) is too limited to satisfy anything other than very elementary boundary conditions. A more general solution follows from linear superposition of elementary solutions (7.2),

$$\Phi(\rho, z) = \frac{1}{2\pi}\int_{-\infty}^{\infty} dK\, \tilde{\Phi}(K) I_0(K\rho)\exp(-ikz) , \qquad (7.3)$$

which suggests that $\Phi(\rho, z)$ and $\tilde{\Phi}(K)I_0(KR)$ are a Fourier-transform pair (i.e., one is the Fourier transform of the other). Therefore we can also write the inverse transform,

$$\tilde{\Phi}(K) = \frac{1}{I_0(K\rho)}\int_{-\infty}^{\infty} dz\, \Phi(\rho, z)\exp(ikz) . \qquad (7.4)$$

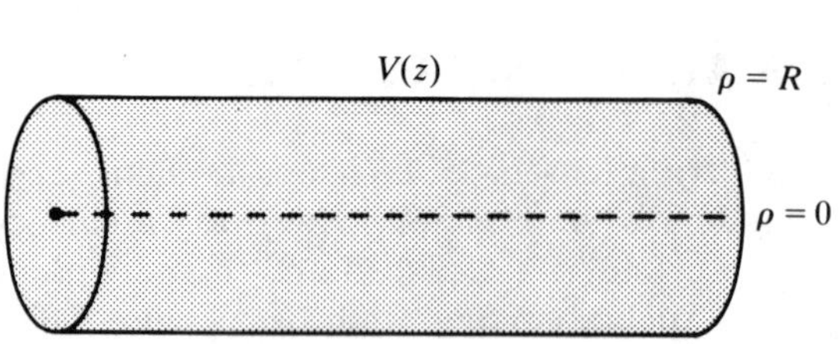

Fig. 7-1. A cylindrically symmetric region bounded by the potential $V(z)$ at $\rho = R$.

Actually, (7.4) defines the Fourier transform of $\Phi(\rho, z)/I_0(K\rho)$ with respect to z, and (7.3) is the inverse transform. However, (7.3) represent a solution of (7.1), and that is the point here. This solution is sufficient to allow us to match boundary conditions to a unique solution. Simply apply (7.4) to the boundary surface $\rho = R$ to obtain

$$\tilde{\Phi}(K) = \frac{1}{I_0(KR)} \int_{-\infty}^{\infty} dz\, V(z) \exp(ikz)\,, \tag{7.5}$$

which yields an explicit expression for the expansion coefficient $\tilde{\Phi}(K)$ because $V(z)$ is our given boundary potential (see above). Insert this into (7.3) to obtain

$$\Phi(\rho, z) = \frac{1}{2\pi} \int_{-\infty}^{\infty} d\zeta\, V(\zeta) \int_{-\infty}^{\infty} dK \cdot I_0(K\rho)\, \frac{\exp[-ik(z-\zeta)]}{I_0(KR)}\,, \tag{7.6}$$

which can be written more conveniently as a convolution,

$$\begin{aligned} \Phi(\rho, z) &= \int_{-\infty}^{\infty} d\zeta\, g(\rho, z-\zeta) V(\zeta)\,, \\ g(\rho, z-\zeta) &= \frac{1}{2\pi} \int_{-\infty}^{\infty} dK\, I_0(K\rho)\, \frac{\exp[-ik(z-\zeta)]}{I_0(KR)}\,. \end{aligned} \tag{7.7}$$

Numerical evaluation of the potential for given $V(z)$ is not entirely trivial because (7.7) requires two quadratures (integrations) to be performed, and one of them is a Fourier transform of a ratio of Bessel functions, i.e., is oscillatory (which usually presents additional difficulties for accuracy). However, the following approximate development represents a considerable simplification at the cost of some accuracy.

We found in Eq. (3.70) that axisymmetric potentials allow us to express $\Phi(\rho, z)$ as a weighted sum of the $2n$th derivatives with respect to z of the axial potential $\Phi_0(z) \equiv \Phi(0, z)$. When that expansion has been made, we need evaluate only

$$g(0, \Delta z) = \frac{1}{2\pi} \int_{-\infty}^{\infty} dK\, \frac{\exp(-ik\,\Delta z)}{I_0(KR)}\,, \tag{7.8}$$

and its derivatives with respect to $\Delta z = z - \zeta$. A remarkable mathematical coincidence found by Bertram (1940), among others, shows that

$$\frac{1}{2\pi}\int_{-\infty}^{\infty} dK\,\frac{\exp(-ik\,\Delta z)}{I_0(KR)} \approx \frac{w_0}{2R}\,\mathrm{sech}^2\!\left(w_0\,\frac{\Delta z}{R}\right), \tag{7.9}$$

provided we choose $w_0 \approx 1.318$ (which is obtained from numerical integration of this equation for $\Delta z = 0$). The error in this equality is of the order of 5% for various values of $\Delta z/R$.

The above model, to which Grivet (1972) also devotes some space, can be extended to two terms,

$$\frac{1}{2\pi}\int_{-\infty}^{\infty} dK\,\frac{\exp(-ik\,\Delta z)}{I_0(KR)} \approx \sum_{i=1}^{2}\frac{w_i}{2R}\,\mathrm{sech}^2\!\left(\bar{w}_i\,\frac{\Delta z}{R}\right), \tag{7.10}$$

with $w_1 \approx 0.963115$, $w_2 \approx 0.363123$, $\bar{w}_1 \approx 1.20241$, and $\bar{w}_2 \approx 1.82466$. These constants are obtained by fitting (7.10) to (7.8) for $\Delta z = 0$ and ∞. The maximum error is then less than 1.5% for $|\Delta z| \leqslant 3$.

In Table 7.1, we compare the one-term approximation (7.9) and the two-term approximation (7.10) to values obtained from (7.6) by direct numerical integration (accurate to three significant figures). It is quite obvious that the two-term approximation is very accurate for nonsmall values of $g(0, \Delta z)$. Let us now return to (7.7) and (7.10) for $\rho = 0$, i.e., for axial values of the potential, and let us integrate by parts to obtain

TABLE 7.1

Δz	Numerical	One-Term Approx.	Two-Term Approx.
0.0	0.663	0.659	0.663
0.1	0.651	0.648	0.650
0.2	0.616	0.615	0.614
0.3	0.563	0.566	0.560
0.4	0.496	0.500	0.505
0.5	0.432	0.439	0.429
0.6	0.365	0.372	0.364
0.7	0.304	0.314	0.303
0.8	0.249	0.254	0.249
0.9	0.203	0.206	0.203
1.0	0.163	0.164	0.164
1.5	0.0515	0.0487	0.056
2.0	0.0156	0.0134	0.0159
2.5	0.00471	0.00361	0.00477
3.0	0.00142	0.00097	0.00143

$$\Phi(0, z) \approx \tfrac{1}{2}(V_i + V_f) + \sum_{i=1}^{2} \frac{w_i}{2\bar{w}_i} \int_{-\infty}^{\infty} d\zeta \, \frac{dV}{d\zeta} \tanh\left[\frac{\bar{w}_i(z-\zeta)}{R}\right], \tag{7.11}$$

given that $V_i \equiv \Phi(0, -\infty)$ and $V_f \equiv \Phi(0, +\infty)$. A significant advantage of this approximate equation (approximate because we used (7.10) for the Green's function $g(0, z - \zeta)$) lies in the fact that the integrand is nonzero only in the gaps between cylinders (where the boundary potential varies). We immediately observe that small gaps imply that the argument of the hyperbolic tangent will not vary appreciably over the range $d\zeta$ between two adjacent cylinders so that some "average" value of $dV/d\zeta$ is essentially what is required. For that reason, we use the linear slope as a sufficient representation of this average.

Equation (7.11) is easily applied to the system of cylinders outlined at the outset of this section. The linear approximation is quite useful because it allows us to integrate the remaining integrand analytically. Consider the sketched region in Fig. 7-2. The region $z_{mi} < z < z_{mf}$ of the mth gap between two adjacent cylinders contributes to (7.11) the terms,

$$\sum_{i=1}^{2} \frac{Rw_i}{2\bar{w}_i^2} \left(\frac{dV}{dz}\right)_m \ln\left\{\frac{\cosh[\bar{w}_i(z - z_{mi})/R]}{\cosh[\bar{w}_i(z - z_{mf})/R]}\right\}, \tag{7.12}$$

and consequently we obtain

$$\Phi(0, z) \approx \tfrac{1}{2}(V_i + V_f) + \sum_{i=1}^{2} \frac{Rw_i}{2\bar{w}_i^2} \sum_m \left(\frac{dV}{dz}\right)_m \ln\left\{\frac{\cosh[\bar{w}_i(z - z_{mi})/R]}{\cosh[\bar{w}_i(z - z_{mf})/R]}\right\}. \tag{7.13}$$

The summation over m should include all the gaps between cylinders: one less than the total number of equi-radius cylinders around the z axis. The final step would be to obtain $\Phi(\rho, z)$ from this by applying (3.70); the derivatives of (7.13) are easily obtained analytically.

An alternative approximation of $\Phi(\rho, z)$ can be obtained in a similar fashion by directly approximating the off-axis Green's function $g(\rho, \Delta z)$.

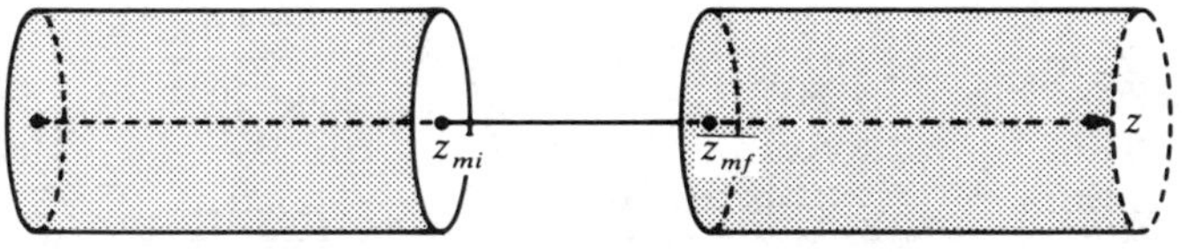

Fig. 7-2. Illustration of the geometry needed for Eq. (7.13).

By means of similar matching techniques to extreme values of (7.8), we find

$$g(\rho, \Delta z) \approx A_1(\rho) \tanh\left(w_1 \frac{\Delta z}{R}\right) + A_2(\rho) \tanh\left(w_2 \frac{\Delta z}{R}\right), \tag{7.14a}$$

with the auxiliary definitions,

$$A_1(\rho) = \sum_{j=1}^{5} a_{1j}\left(\frac{\rho^2}{R^2}\right)^{j-1}, \qquad A_2(\rho) = 1 - A_1(\rho),$$
$$w_1 = 1.20241, \qquad w_2(\rho) = \sum_{j=1}^{5} \bar{w}_{2j}\left(\frac{\rho^2}{R^2}\right)^{j-1}. \tag{7.14b}$$

The coefficients that appear in (7.14b) are given in Table 7.2 to six significant figures.

The off-axis potential is then estimated reasonably well by the expression,

$$\Phi(r, z) \approx \tfrac{1}{2}(V_i + V_f) + \frac{R}{2} \sum_{m=1}^{N-1} \left(\frac{dV}{d\zeta}\right)_m \sum_{i=1}^{2} \frac{A_i(\rho)}{w_i(\rho)} \ln\left\{\frac{\cosh[w_i(z - z_{mi})/R]}{\cosh[w_i(z - z_{mf})/R]}\right\}, \tag{7.15}$$

at least for values of ρ/R that are not beyond 0.5. Note that the coefficients $\bar{w}_{2j}$ in Table 7.2 pertain to the coefficient $w_2(\rho)$ in (7.14), and that the $w_2(\rho)$ in (7.15) depends upon ρ even though argument ρ is not shown explicitly in the hyperbolic cosine functions. There are some minor advantages in using (7.15) rather than derivatives of (7.13) in an off-axis expansion of $\Phi(\rho, z)$. Table 7.3 shows a comparison of exit ray heights ρ and ray tangents ρ' for 21 electron trajectories entering a two-cylinder

TABLE 7.2

	$j=1$	$j=2$	$j=3$	$j=4$	$j=5$
a_{1j}	0.800987	−1.15806	0.418581	0.0672094	0.00587265
$\bar{w}_{2j}$	1.82462	2.61829	−6.18007	26.2039	−33.9145
b_{2j}	0.109074	0.476788	−0.476736	−0.279007	0.363195

TABLE 7.3

	Numerical		Eq. (7.13)		Eq. (7.15)	
Ray #	ρ	ρ'	ρ	ρ'	ρ	ρ'
0p	−2.65	0.30	−2.68	0.21	−2.75	0.19
1d	6.06	0.50	6.15	0.71	6.21	0.68
1p	0.47	0.17	0.48	0.19	0.48	0.19
1u	1.69	1.08	1.72	1.14	1.72	1.13
2d	8.93	1.19	9.06	1.49	9.06	1.35
2p	5.62	1.20	5.71	1.40	5.77	1.38
2u	6.45	1.20	6.56	2.22	6.60	2.19
3d	11.93	1.67	12.12	2.11	11.99	1.78
3p	10.20	1.98	10.35	2.34	10.31	2.15
3u	11.32	2.73	11.49	3.12	11.41	2.89
4d	14.67	2.03	14.88	2.53	14.59	1.98
4p	15.27	2.60	15.49	3.12	15.18	2.55
4u	16.56	3.18	16.81	3.76	16.44	3.11
5d	19.16	2.21	19.43	2.86	18.86	1.91
5p	20.00	2.78	20.28	3.42	19.68	2.44
5u	22.54	3.07	22.88	3.89	22.16	2.71
6d	23.00	1.78	23.33	2.56	22.54	1.26
6p	24.90	2.20	25.25	3.04	24.38	1.54
6u	28.64	2.00	29.02	2.91	27.84	0.60
7d	25.86	0.16	26.23	1.09	25.23	−0.69
7p	29.23	0.36	29.63	1.35	28.27	−1.23
Units	mil	mrad	mil	mrad	mil	mrad

lens (one gap) with a 25:6 voltage ratio on the cylinders and with gap width equal to $\frac{1}{2}R$. This is a rather large gap width, and it was chosen to show that, even then, some rather useful results can be obtained from the above analytical approximations.

The columns in Table 7.3 are marked with reference to the equation from which the numbers have been derived. Columns 2 and 3 pertain to computer numerical simulations of the trajectories by means of finite-mesh relaxation methods (see Section 4.1.5.1). The notations in the first column are intended to indicate that the trajectories were produced at a cathode (at a location prior to the first cylinder) at seven equally spaced intervals in groups of three: a principal (e.g., 3p), a 45° up electron (3u), and a 45° down electron (3d). By "45° up" we mean that the initial velocity vector of the electron makes an angle of +45° with the horizon-

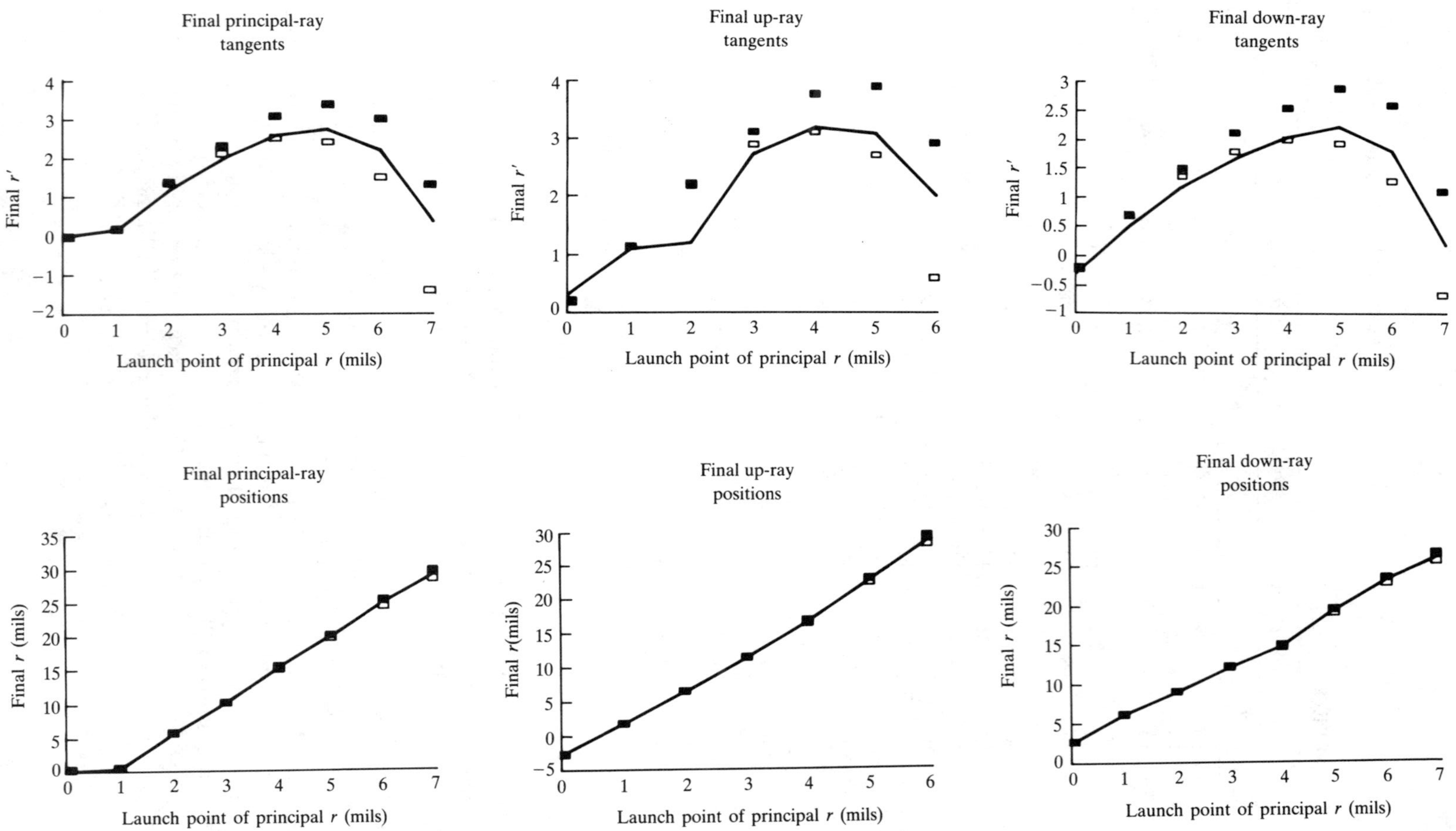

Fig. 7-3. A series of graphs of the $\rho(\rho_0)$ and $\rho'(\rho_0)$ values in Table 7.3 for various one-term and two-term approximations of the potentials inside compound cylindrical lenses. In the graph ■ is old 2-term, □ is new 2-term, and — is num r' (Top) and num r (bottom).

tal. The interesting feature here is that the approximations (7.13) and (7.15) have been used for the potential, and these potentials have *then* been inserted into trajectory equations, which finally leads to the ρ and ρ' numbers in Table 7.3. Thus whatever errors there are in potential will accumulate in the final trajectory parameters tabulated in Table 7.3. Clearly, the trajectory locations approximated reasonably well, and apparently the ray tangents deviate from the numerically accurate values, especially at the lower end of the table for electrons that are launched further away from the axis. These data are also summarized in Fig. 7-3. The fact that the trajectory locations are, and the ray tangents are not, accurate indicates only that the endpoint (which is well into the field-free region inside the second cylinder) appears to have been chosen such that errors are just beginning to accumulate beyond the degree of reasonable tolerance. The above example is clearly a very severe test of the approximations.

Cylinder lenses are used in electron guns to focus a beam that has been formed with a prior set of aperture lenses (see Chapter 11). *Bipotential* lenses consist of a pair of successive cylinders in which the second is at a higher voltage than the first. *Unipotential* (or *einzel*) lenses consist of at least three cylinders with the first and last at the same voltage so that the electron velocity is unchanged by the lens. All kinds of combinations and extensions are possible, and more accurate performance characteristics need to be evaluated numerically (Harting and Read 1976). The approximate analysis, however, serves to give a preliminary idea of what to expect from a new design, but the analysis of this section is limited, of course, by the requirement that all cylinders are circular and that they have one common radius.

7.2 ELECTROSTATIC APERTURE LENSES

Another category of cylindrically symmetric lenses is formed from circular apertures (holes) in planar metal plates, e.g., as sketched in Fig. 7-4 in the presence of two electric fields imposed by sources elsewhere. Let the metal plate be at the fixed voltage V_0, and let the asymptotic fields be $\vec{E}_L$ at $z<0$ and $\vec{E}_R$ at $z>0$. These two fields are assumed to be uniform asymptotically far away from the plate, but of course they are modified by the presence of a metal plate at voltage V_0 with a circular aperture. We adopt the convention that the $\vec{E}$ fields are as sketched in Fig. 7-4, i.e., that the potential $\Phi(\rho, z)$ increases as z increases.

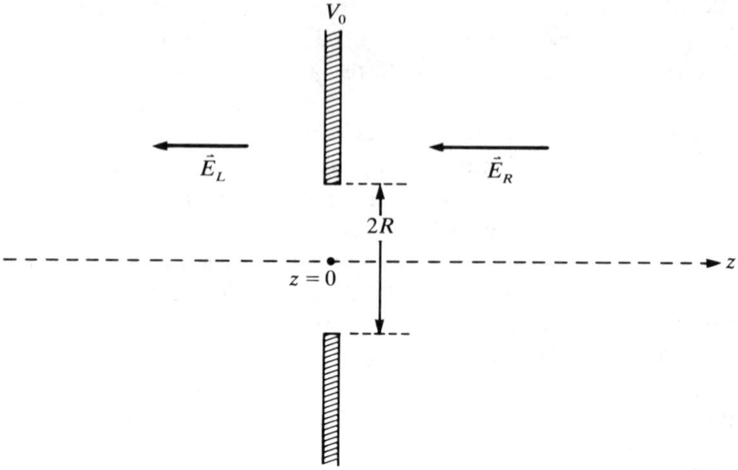

Fig. 7-4. The basic electrostatic aperture lens inserted into a constant field region.

The analysis of this type of lens appears first to have been done accurately by *Regenstreif* (1951). He used confocal elliptical and hyperbolic coordinates to derive a rigorous expression for the potential $\Phi(\rho, z)$; the derivation can be found in a recent text by *Szilagyi* (1988). As it is lengthy, and as it is a straightforward application of a separation-of-variables technique in the chosen coordinate system, we will merely write down the result,

$$\Phi(\rho, z) = V_0 + \tfrac{1}{2}(E_L + E_R)z - \frac{1}{\pi}(E_L - E_R)|z|\left\{\frac{R}{\mu} + \tan^{-1}\left(\frac{\mu}{R}\right)\right\},$$

$$\mu(\rho, z) = \sqrt{\tfrac{1}{2}(\rho^2 + z^2 - R^2) + \tfrac{1}{2}\sqrt{(\rho^2 + z^2 - R^2)^2 + 4z^2R^2}}\,. \qquad (7.16)$$

The convention for the $\vec{E}$ fields is needed to define their magnitudes in the above expressions. The magnitude of $\vec{E}_L$ or $\vec{E}_R$ is considered to be negative if either field is reversed in direction from what was sketched in Fig. 7-4, and that would change some of the signs in (7.16).

Let us illustrate the application of this expression for the potential by considering some limiting cases.

1. Far from the axis ($\rho \gg R$):

The quantity $\mu(\rho)$ tends towards ρ as $\rho \to \infty$, with errors of $O(z/\rho)$ and $O(R/\rho)$. Hence $R/\mu \to 0$ and $\tan^{-1}(\mu/R) \to \pi/2$, and we obtain

$$\Phi(\rho, z) \approx V_0 + E_R z \quad \text{for } z > 0\,,$$
$$\approx V_0 + E_L z \quad \text{for } z < 0\,. \tag{7.17}$$

This is precisely what is expected when the plate has no aperture, and the condition $\rho \gg R$ indicates that the location is sufficiently distant from the aperture (provided also that $z \ll \rho$) to approximate that situation.

2. On the axis ($\rho = 0$):

The quantity $\mu(0) = |z|$ on the axis, and consequently (7.16) reduces to

$$\Phi(\rho, z) = V_0 + \tfrac{1}{2}(E_L + E_R)z - \frac{1}{\pi}(E_L - E_R)R\left\{1 + \frac{z}{R}\tan^{-1}\left(\frac{z}{R}\right)\right\}. \tag{7.18}$$

At $z = 0$, i.e., at the center of the aperture, the axial potential is $\Phi(0,0) = V_0 + (R/\pi)(E_L - E_R)$ so that $\Phi(0,0) > V_0$ when $E_L < E_R$, and vice versa. We have the four possible situations depicted in Fig. 7-5, which cover all possible directions of the fields $\vec{E}_L$, $\vec{E}_R$ with electron motion either from left to right or vice versa. Situations (a) and (b) of Fig. 7-5 represent the two more interesting ones when electrons are incident from the left. The former represents a divergent, and the latter a convergent lens. Situation (d) has some analogy to a unipotential lens: electrons are decelerated initially in a convergent-lens region, and they are either "reflected" (because their initial kinetic energy is insufficient to overcome the cumulative effect of E_L) or they undergo a converging force in a region where they move more slowly before being accelerated again by E_R. Hence the entire lens is either an electron mirror or it is a convergent lens. Situation (c) is known as an *electronic aperture* because a decrease in voltage V_0 will "raise" the barrier of low potential at the aperture and thus "choke off" the passasge of electrons beyond it.

One might see a contradiction in the situation (a) in which the potential increases steadily from left to right, but the lens is *divergent*, in apparent contradiction of a statement we made in Chapter 6 to the effect that accelerating lenses always focus! That statement, however, is restricted to those lenses that are characterized by entry and exit regions at *constant* (albeit different) potentials. Here, we have lenses with constant entry and exit *fields*.

Even though Eq. (7.16) allows accurate numerical calculation of electron trajectories to be made, it is sometimes useful to have a

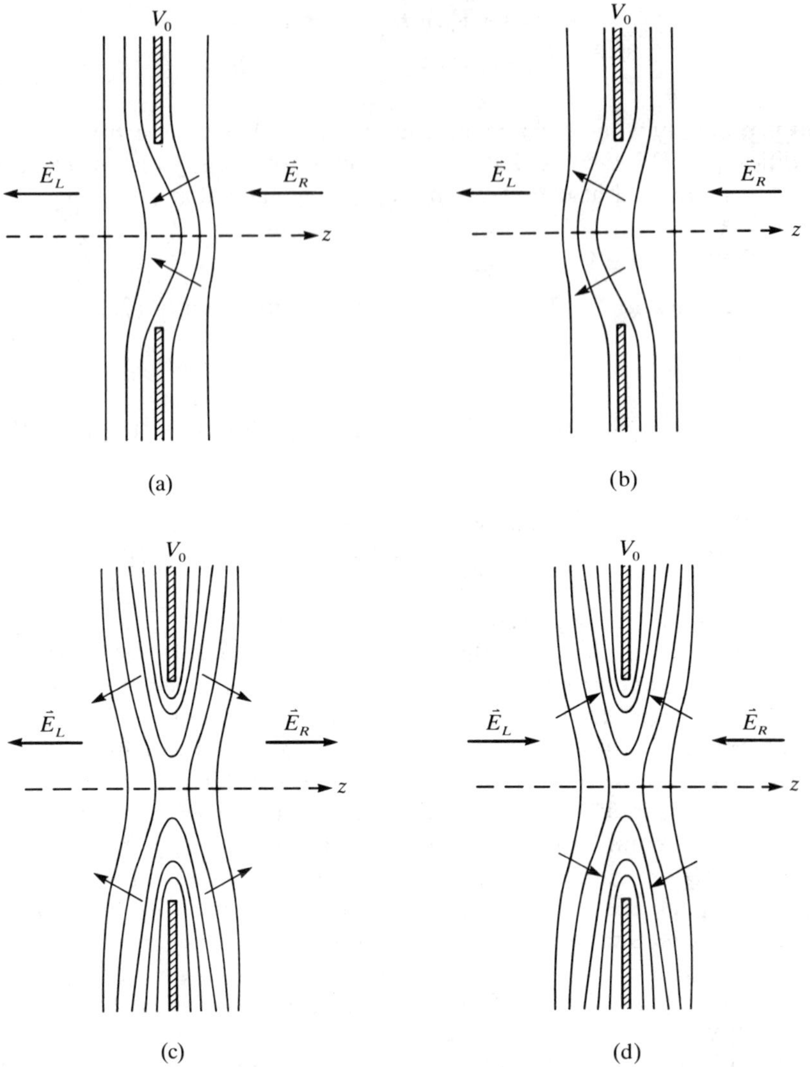

Fig. 7-5. Equipotential surfaces for electrostatic aperture lenses inserted into constant-field regions for four different sets of asymptotic fields.

"quick-and-dirty" approximate analytical estimate of the lens properties. Such an estimate preceded the Regenstreif calculation historically; it appears to have been made first by Davisson and Calbick (1931). They made an analogy to a thin-lens approximation: the region in which the $\vec{E}$ fields are not constant is restricted to $-z_0 > z < +z_0$, where $z_0 \ll R$. The situation is illustrated in Fig. 7-6 for a convergent lens ($E_R > E_L$; both fields are accelerating). Upon applying the paraxial formula (6.15), we observe that

$$\sqrt{\Phi_0(z_2)}\rho'(z_2) - \sqrt{\Phi_0(z_1)}\rho'(z_1) = -\frac{1}{4}\int_{-z_0}^{z_0} d\zeta \, \frac{\Phi_0''(\zeta)}{\sqrt{[\Phi_0(\zeta)]}} \rho(\zeta) \,. \tag{7.19}$$

We make the thin-lens approximations,

$$\rho(\zeta) \approx \rho(0) \quad \text{and} \quad \Phi_0(\zeta) \approx V_0 \quad \text{for } -z_0 < \zeta < +z \,, \tag{7.20}$$

to convert the integral in (7.19) into

$$\int_{-z_0}^{z_0} d\zeta \, \frac{\Phi_0''(\zeta)}{\sqrt{[\Phi_0(\zeta)]}} \rho(\zeta) \approx \frac{\rho(0)}{\sqrt{V_0}} \int_{-z_0}^{z_0} d\zeta \, \Phi''(\zeta) \approx -\frac{\rho(0)}{\sqrt{V_0}} (E_R - E_L) \,. \tag{7.21}$$

At the location z_2 (see Fig. 7-5), we have $\rho'(z_2) = -\tan\theta$, and $\sqrt{\Phi_0(z_2)} = V_0 + E_R z_2 \approx V_0$ so that we then obtain

$$\tan\theta \approx \rho(0) \, \frac{E_R - E_L}{4V_0} \,. \tag{7.22}$$

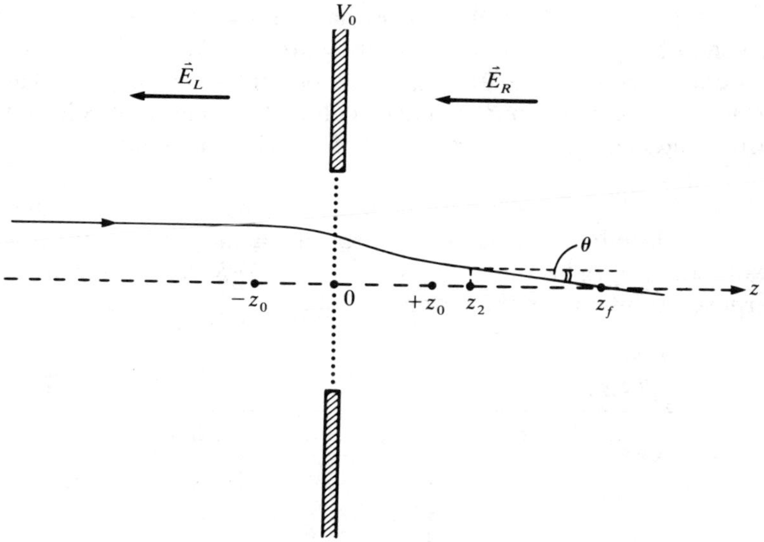

Fig. 7-6. A principal-electron trajectory through an aperture lens.

This is obviously not correct for large z_2 because the $\vec{E}_R$ field continues to accelerate the electron in the z direction and thus to decrease the angle θ. However, if the E_R field is weak or zero, then (7.22) might serve as a rough estimate. In the latter case of zero E_R field, then $\Phi(z_f) \approx V_0$ at any final location $z_f = f$ on the axis, and we may set $\tan\theta = -\rho(0)/f$ to obtain a focal-length estimate,

$$f \approx \frac{4V_0}{E_L - E_R} \, . \tag{7.23}$$

The lens is convergent when $E_R \sim 0$ as f is then positive. The usefulness of these estimates has probably diminished in the present age of easy access to fast computers; numerical calculations based upon (7.16) will approximate the precise focal length to any numerically desirable accuracy.

7.3 COMPOUND APERTURE LENSES

The exact solution of the potential field of a single aperture lens forms the basis for approximating the solution for a compound lens consisting of two or more aperture lenses. An example of such a lens is sketched in Fig. 7-7; it corresponds to the type of electron lens that is used in cathode-ray tubes (CRTs) to extract a beam of electrons from a planar cathode that emits electrons (somewhat "indiscriminately") in a very wide cone of angles over a large emitting area. If the cathode is at $z = 0$, then we shall define $\Phi(\rho, 0) \equiv V_0 = 0$ as our reference voltage. The two thick apertures can be approximated by four thin ones that coincide with the front and back surfaces, and of which we give the parameters in Table 7.4.

We refer to Fig. 7-7 again for definition of the locations and the other parameters listed in the table. We specify a uniform field $\vec{E}_4$ asymptotically to the right of $z = z_4$ (i.e., for $\rho \gg R_2$). We try a linearly superposed solution for the potential field,

TABLE 7.4

Aperture #	Location	Radius	Voltage
1	z_1	R_1	V_1
2	z_2	R_1	V_1
3	z_3	R_2	V_2
4	z_4	R_2	V_2

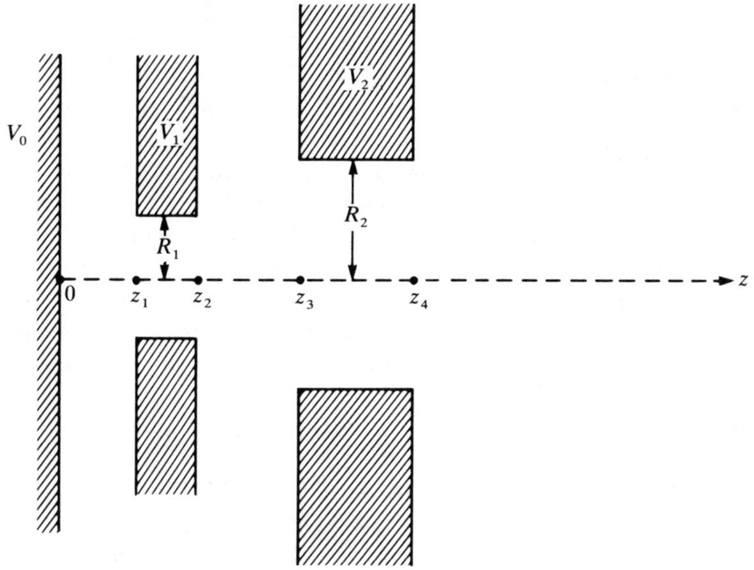

Fig. 7-7. Modeling of the beamforming region of an electron gun by compound aperture lenses.

$$\Phi(\rho, z) = \sum_{m=1}^{4} \Phi_m(\rho, z) ,$$

$$\Phi_m(\rho, z) = \frac{1}{\pi} E_m \left\{ |z + z_m| \left[\frac{1}{u_m^+} + \tan^{-1}(u_m^+) \right] - |z - z_m| \left[\frac{1}{u_m^-} + \tan^{-1}(u_m^-) \right] \right\} , \tag{7.24a}$$

with the auxiliary definitions,

$$(u_m^{\pm})^2 \equiv \frac{1}{2R_m} \{ (r^{\pm})^2 + \sqrt{(r^{\pm})^4 + 4R_m^2(z \pm z_m)^2} \} ,$$

$$(r^{\pm})^2 \equiv r^2 + (z \pm z_m)^2 - R^2 , \tag{7.24b}$$

$$E_1 = \frac{V_1 - V_0}{z_1} , \qquad E_2 = -E_3 = \frac{V_2 - V_1}{z_2 - z_1} .$$

Actually there are four more apertures, so that we obtain a total of *eight*. We introduce four mirror images at $z = -z_m$ ($m = 1$, 2, 3, and 4) with voltages $-V_m$ (see Table 7.4) on their plates. These four have contributions similar to (7.24a), and they are needed to make $V_0 = 0$ at $z = 0$. The linear superposition is not an exact solution but there are regions where it is a close approximation. The regions where it is not a good approximation are the immediate environment of the inside surfaces of the two thick apertures, i.e., the regions $z_1 < z < z_2$ with $\rho \approx R_1$ and $z_3 < z < z_4$ with $\rho \approx R_2$. The potential close to the axis is approximated reasonably well when $|z_{i+1} - z_i| \ll R_i$ (see de Wolf 1982). The superposition is a generalization of one only for the axial potential of three apertures proposed by Regenstreif (1951). Recent work (de Wolf 1982) discusses some tests of the approximation for the simulated beamforming regions of electron guns used in TV kinescopes, and electron trajectories calculated with this approximate potential field show remarkable agreement with numerically accurate computer simulations.

7.4 STRONG-FOCUSSING (QUADRUPOLE) LENSES

All electrostatic lenses discussed up to this point produce focussing or defocussing by means of an electrostatic force with a component $\vec{E}_T$ in a direction perpendicular to the axis, and with $|E_T| \ll |E_z|$. Cylinder lenses in particular are relatively weak for the reason illustrated in Fig. 7-8: electrons entering from the lower-voltage (V_1) side are first focussed at lower velocities and then defocussed as they accelerate into the V_2 region. The net effect is *focussing* because the electrons spend less time in the defocussing region (where the v_z component of the velocity is higher). To state this more quantitatively we apply the focal-distance formulas (6.19) to a principal electron (i.e., an electron with initial velocity parallel to the z axis):

$$\frac{1}{f} = \frac{1}{4\sqrt{V_2}} \int_{-\infty}^{\infty} d\zeta \, \frac{\Phi''(0, \zeta)}{\sqrt{[\Phi(0, \zeta d)]}} \, \frac{\rho(\zeta)}{\rho(-\infty)} \, . \tag{7.25}$$

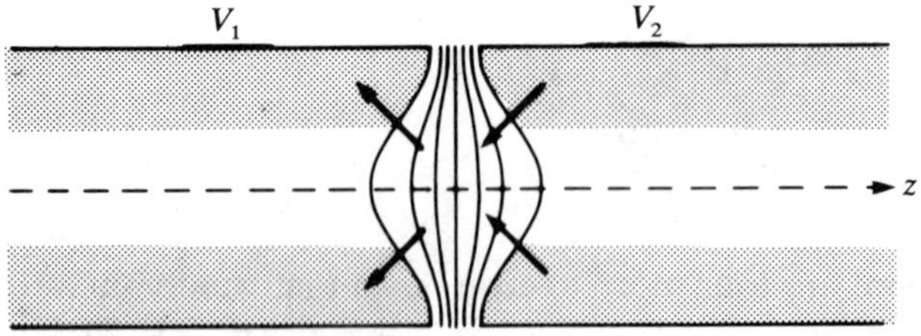

Fig. 7-8. Equipotential surfaces in a typical weak-focussing lens.

The axial potential $\Phi_0(z)$ of the lens portrayed in Fig. 7-8 is sketched in Fig. 7-9. Initially, focal distance f would be decreased as we move the upper bound of the integral in (7.25) through the $\Phi_0'' > 0$ region, but this is followed by an increase in f as the upper bound moves through the $\Phi_0'' < 0$ region (in which $\rho(z)/\sqrt{\Phi_0(z)}$ is smaller so that a net positive f ensues). These counterbalancing effects therefore produce relatively weak focussing action.

This effect appears to be overcome in aperture lenses; at least Fig. 7-5b shows no $\Phi_0'' < 0$ region. Nevertheless, aperture lenses are relatively weak in any realistic situation due to the small ratio of $|E_T|/|E_z|$ inherent in the structure of the electrostatic field produced by such lenses. Of course, focussing strengths can be increased by design attempts to increase $\Phi_0''(z)/\sqrt{\Phi_0(z)}$ in the lensing region, but this requires very high voltage ratios V_2/V_1 over short gaps, i.e., danger of electric breakdown, in cylinder lenses. The practicality of doing so is somewhat limited. A further discussion of limitations of cylinder and aperture lenses is given by Grivet (1972, Chapter 10).

The problem with the class of lenses that are governed, at least paraxially, by Eqs. (6.21) can also be formulated by noting that both of these equations contain the focussing factor Φ_0'/Φ_0 to the second power. That fact is immediately obvious for the Picht formulation in which the factor Φ_0'/Φ appears quadratically, but it is also there in the first of Eqs. (6.21). Integrate that equation in parts from $z = -\infty$ to the inflection point (at least *one* must occur), and then to the next one, and finally from that to $z = \infty$. If the lens is sufficiently weak so that we may approximate $\rho(z) \approx \rho_L$, we will obtain a factor $(\Phi_0')^2/(2\Phi_0^{3/2})$ in the integrand. If (Φ_0'/Φ_0) is a small ratio, then this factor appears at least quadratically and the focussing factor is a *second-order effect* (Lawson 1988).

A philosophy of designing strong-focussing lenses is based upon at-

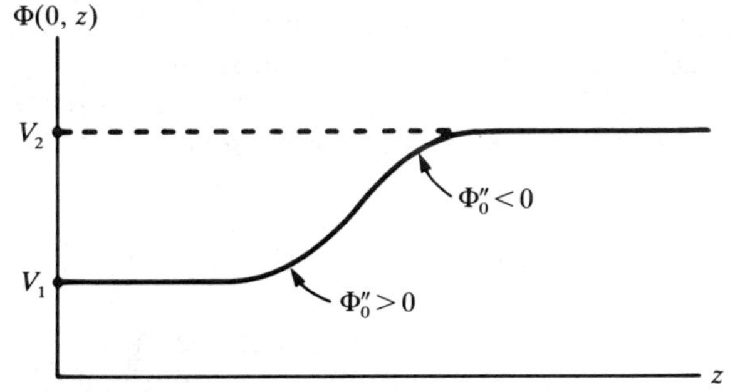

Fig. 7-9. The axial potential profile of the lens of Fig. 7-8.

tempts either to increase the focussing factor or to increase the ratio of $|E_T|/|E_z|$. It is perhaps somewhat more direct to attempt the latter. Let us consider an electrostatic potential $\Phi(x, y, z)$ with as much lensing action in the transverse and as little in the longitudinal direction as possible. Let $\Phi_0(z)$ be the axial potential. By virtue of a Taylor expansion, we obtain

$$\Phi(x, y, z) = x \frac{\partial \Phi_0}{\phi x} + y \frac{\partial \Phi_0}{\partial y} + \frac{1}{2}\left[x^2 \frac{\partial^2 \Phi_0}{\partial x^2} + y^2 \frac{\partial^2 \Phi_0}{\partial y^2} + 2xy \frac{\partial^2 \Phi}{\partial x \partial y}\right] + \cdots, \tag{7.26}$$

where the notation $\partial \Phi_0/\partial x$ is an abbreviation for $\partial \Phi/\partial x$ evaluated on the axis, and similarly for the other derivatives. We shall require the symmetry properties $\partial \phi_0/\partial x = 0$ and $\partial \phi_0/\partial y = 0$. This is implied by requiring the slightly stronger conditions $\Phi(-x, y, z) = \Phi(x, y, z)$ and $\Phi(x, -y, z) = \Phi(x, y, z)$. This is a necessary but not sufficient condition for cylindrical symmetry, but we shall not require axial symmetry. The condition also force $\partial^2 \Phi_0/\partial x\, \partial y = 0$. If we use that, and the additional terms that we obtain by eliminating $\partial^2 \Phi_0/\partial x^2$ with Laplace's equation, we obtain

$$\Phi(x, y, z) = \Phi_0(z) - \tfrac{1}{2}(x^2 - y^2) \frac{\partial^2 \Phi_0}{\partial y^2} - \tfrac{1}{2} x^2 \frac{\partial^2 \Phi_0}{\partial z^2} + \cdots, \tag{7.27}$$

where the ellipsis indicates terms of at least fourth-order in x and/or y. It is obvious from (7.25) that $\partial^2 \Phi_0/\partial z^2$ cannot be zero everywhere if we have a lens, but we can have lensing action in a region where this quantity is zero, e.g., as at a point of inflection (as in Fig. 7-9). Let us assume a lensing region in which Φ_0 is independent of z. There would be zero longitudinal force there. Metal structures that show no change of shape in the z direction would allow potentials with that property (although, obviously, infinitely long structures are impossible). In such a region, we have

$$\Phi(x, y, z) = \Phi_0(z) - \tfrac{1}{2}\kappa^2(x^2 - y^2) + O(x^4, y^4, x^2y^2), \tag{7.28}$$

where $\kappa \equiv \partial^2 \Phi_0/\partial y^2$. Electrostatic lenses that obey (7.28) are known as *quadrupole lenses*, and they can be formed from four long pieces of metal with cross-sectional hyperbolic shape so that the location of each obeys the equation $x^2 - y^2 = \text{constant}$ (see Fig. 7.10), which also shows the defining voltages. The solution of the two-dimensional Laplaces's equation that satisfies the boundary conditions posed by this structure is $\Phi(x, y) = \Phi_0 - \frac{1}{2}\kappa(x^2 - y^2)$. The boundary conditions imply

$$\Phi(\pm a, 0) = \Phi_0 - \tfrac{1}{2}\kappa^2 a^2\,, \qquad \Phi(0, \pm b) = \Phi_0 - \tfrac{1}{2}\kappa^2 b^2\,. \tag{7.29}$$

Because $\Phi(\pm a, 0) = V_1$ and $\Phi(0, \pm b) = V_2$, it follows that

$$V_1 = \Phi_0 - \tfrac{1}{2}\kappa^2 a^2\,, \qquad V_2 = \Phi_0 + \tfrac{1}{2}\kappa^2 b^2\,, \tag{7.30}$$

suffices to create a "perfect" lens that satisfies (7.28); "perfect" because there are no higher-order terms. Of course in actual fact there has to be z-dependence of Φ somewhere (see below). If that fact is ignored temporarily, then inversion of (7.30) yields

$$\kappa^2 = \frac{2(V_2 - V_1)}{a^2 + b^2}\,, \tag{7.31}$$

hence the strength factor κ can be controlled by choice of $V_2 - V_1$ and the diameters $2a$, $2b$. The metallic plates are located at $x^2 - y^2 = a^2$ and at $y^2 - x^2 = b^2$. The electrostatic fields are easily obtained from $\vec{E} = -\check{\nabla}\Phi$ in (7.28), and they are

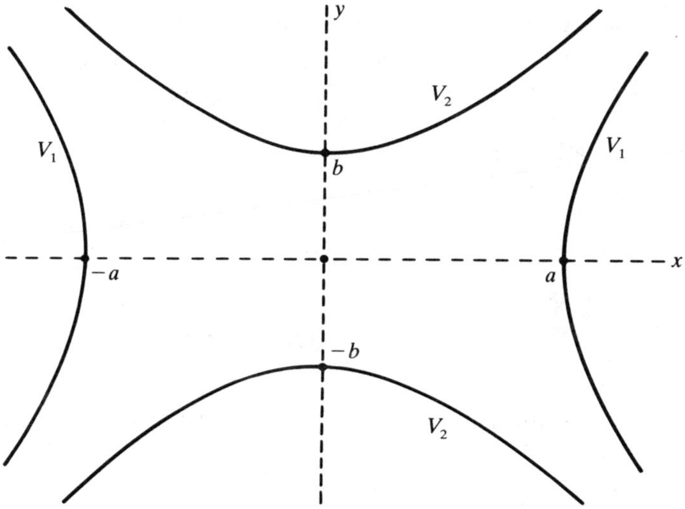

Fig. 7-10. A quadrupole lens formed by four hyperbolic bounding surfaces.

$$E_x = \kappa^2 x\,, \qquad E_y = -\kappa^2 y\,. \tag{7.32}$$

We see that there is strong focussing in the plane $x = 0$, but it goes at the cost of *defocussing* in the $y = 0$ plane. This is not necessarily a problem in some applications. In TV kinescopes, for example, the phosphors that must be illuminated by a focussed electron beam often lie in stripes inside the screen. There may be a need to focus strongly in a direction that is parallel to the stripes; the width of the stripes may suffice in the perpendicular direction. Alternatively, a *shadow mask* may be used to make sure that only certain parts of phosphor stripes are illuminated, and the separation of holes in the shadow mask in one dimension may be more critical than in the other; hence strong focussing in that direction may be needed.

In practice, the above idealized quadrupole lens is impossible because the lack of a z-dependence in $\Phi(x, y)$ requires an infinitely long structure in the z direction. Actual lenses are finite in extent. Let us assign a length Δz to such a lens so that the quadrupole lensing following from (7.28) will be restricted to a region $-\frac{1}{2}\Delta z < z < \frac{1}{2}\Delta z$ (if we ignore fringing effects). An estimate of a focal length can then be made, using Eq. (3.18), because we then have in the $x = 0$ plane,

$$y'_f - y'_i \approx -\tfrac{1}{2}\int_{-\Delta z/2}^{\Delta z/2} dz\,\frac{E_y(z)}{\Phi(z)}\,, \tag{7.33}$$

provided $z_f \gg \frac{1}{2}\Delta z$ and $z_i \ll \frac{1}{2}\Delta z$. In applying (3.18) we have assumed no magnetic flux density and also that $E_z \approx 0$ in the lensing region. Furthermore, we have assumed the paraxial approximation $(\vec{\rho}\,')^2 \ll 1$, and also that $y'_i = 0$ by restricting ourselves to a principal electron. If Δz is sufficiently small, then $y(z) \approx y_i$ inside the focussing region. Since $E_y(z) = -\kappa^2 y(z) \approx -\kappa^2 y_i$ in the focussing region, we can find an approximate expression of use for the focal length, provided that $\frac{1}{2}\kappa^2(x_i^2 - y_i^2) \ll \Phi_0$:

$$\frac{1}{f_y} \equiv \frac{y'_f}{y_i} \approx \frac{\kappa^2}{2\Phi_0}\,\Delta z\,. \tag{7.34}$$

Or, in terms of the quadrupole model of hyperbolic cross-sectioned metallic plates,

$$\frac{1}{f_y} \approx \frac{V_2 - V_1}{\Phi_0}\,\frac{\Delta z}{a^2 + b^2}\,. \tag{7.35}$$

We need to remember that the axial potential value Φ_0 (or κ) is not known a priori; we need to calculate it from a Laplace equation determined by the boundary values. In the special case that $a = b$ we may estimate $\Phi_0 \approx \frac{1}{2}(V_1 + V_2)$, provided that $|V_2 - V_1| \ll \Phi_0$, and thus obtain

$$\frac{1}{f_y} \approx \frac{V_2 - V_1}{V_2 + V_1} \frac{\Delta z}{a^2} . \tag{7.36}$$

The estimates based upon (7.33) are all approximate and they serve merely as an order-of-magnitude estimate—perhaps somewhat better than that—of the focal distance.

REFERENCES

Abramowitz, M., and I. A. Stegun (1964) *Handbook of Mathematical Functions*, National Bureau of Standards, Appl. Math. Series 55, U.S. Department of Commerce.

Bertram, S. (1940) *Proc. I.R.E.*, **28**, 418.

Davisson, C. J. and C. J. Calbick (1931) *Phys. Rev.*, **38**, 585.

de Wolf, D. A. (1982) *RCA Rev.*, **43**, 339.

Grivet, P. (1972) *Electron Optics*, Pergamon, Oxford.

Harting, E. and F. H. Read (1976) *Electrostatic Lenses*, Elsevier, Amsterdam.

Lawson, J. D (1988) *The Physics of Charged-Particle Beams*, Clarendon Press, Oxford.

Regenstreif, E. (1951) *Ann. Radioél.*, **6**, 51, 164.

Szilagyi, M. (1988) *Electron and Ion Optics*, Plenum Press, New York.

CHAPTER 8

ELEMENTS OF ABERRATION THEORY

We discussed the lowest-order approximation to the trajectory equations of charged particles in Chapter 5. This approximation gave rise to *paraxial* (or *Gaussian*) optics: essentially a linearized version of the trajectory equations of Chapter 3. It requires small deflection angles and small deviations from the reference trajectory that goes through the central region of an electron optics lens.

Successive corrections to Gaussian optics are obtained by systematic inclusion into the trajectory equations of terms of higher order in the deflection distance ρ and the ray tangent ρ'. The effects of the higher-order terms are known as *aberrations* because they express the deviations, among other things from perfect focus and from other idealized Gaussian-image optics. The study of these aberrations easily can become quite complicated. Many texts give lengthy (if not complete) treatment of the subject (Grivet 1972, El-Kareh and El-Kareh 1970, Glaser 1952, Zworykin et al. 1945). Almost all texts on electron optics devote some space to the subject. We shall not give a lengthy treatment, but we shall try to make the principles as clear as possible. We will follow a line of reasoning that is essentially the same as that given by Grivet (1972) for third-order aberrations; first for axisymmetric lenses in Section 8.1 and then for the more general electrostatic case in Section 8.2. There is little need to give a "complete" treatment for several reasons. First, aberrations of order higher than third hardly occur by themselves because the regime in which fifth-order aberrations are not perturbed by seventhorder ones, etc., is only a small stretch beyond that in which third-order aberrations become appreciable. As in many forms of perturbation analysis, only the first order in perturbations (i.e., third-order aberrations here) is useful. In the second place, it is almost commonplace at this time to simulate particle trajectories by computer calculations with electromagnetic fields calculated numerically with appreciable accuracy from Laplace's or Poisson's equation. The dependence of $\vec{\rho}(z)$ upon $\vec{\rho}(0)$ can be computed numerically and the concomitant aberrations isolated accordingly without deriving page-long aberration formulas for special geometries. Hence it seems useful to restrict ourselves only to the main elements of aberration theory.

8.1 AXISYMMETRIC ELECTROSTATIC LENSES

Let us return to the derivation of trajectory equations in Chapter 5 for paraxial optics, but let us carry the approximations one step further to obtain equations that include third-order terms in ρ and ρ' for axisymmetric fields. The nonrelativistic equation of motion for *meridional* electrons (i.e., electrons with initial velocity vector in one and the same plane with the z axis) in an axisymmetric electrostatic field is (Eq. (3.19)),

$$\rho''(z) = \frac{1+\rho'^2(z)}{2\Phi_a(\rho, z)}\left[\frac{\partial\Phi(\rho, z)}{\partial\rho} - \rho'(z)\,\frac{\partial\Phi(\rho, z)}{\partial z}\right]. \tag{8.1}$$

Apply (3.70), but include terms through fourth order in $\rho(z)$, and consider derivatives,

$$\begin{aligned}
\Phi(\rho, z) &= \Phi_0(z) - \frac{1}{4}\,\Phi_0''(z)\rho^2 + \frac{1}{64}\,\Phi_0^{(4)}(z)\rho^4 - \cdots,\\
\frac{\partial\Phi(\rho, z)}{\partial\rho} &= -\frac{1}{2}\,\Phi_0''(z)\rho + \frac{1}{16}\,\Phi_0^{(4)}(z)\rho^3 - \cdots,\\
\frac{\partial\Phi(\rho, z)}{\partial z} &= \Phi_0'(z) - \frac{1}{4}\,\Phi_0'''(z)\rho^2 + \cdots.
\end{aligned} \tag{8.2}$$

Hence, for small values of $\rho(z)$ we obtain

$$\frac{1}{\Phi_a(\rho, z)} \approx \frac{1}{\Phi_{a0}(z)}\left(1 + \frac{1}{4}\,\frac{\Phi_0''}{\Phi_{a0}}\,\rho^2 - \cdots\right). \tag{8.3}$$

Now insert (8.2) and (8.3) into (8.1) and retain terms of order no higher than third in $\rho(z)$,

$$\rho''(z) = \frac{1+\rho'^2}{2\Phi_{a0}}\left(1 + \frac{1}{4}\frac{\Phi_0''}{\Phi_{a0}}\rho^2\right)$$
$$\times\left(-\frac{1}{2}\Phi_0''\rho + \frac{1}{16}\Phi_0^{(4)}\rho^3 - \Phi_0'\rho' + \frac{1}{4}\Phi_0'''\rho^2\rho'\right), \qquad (8.4a)$$

which can be rearranged to yield an equation with terms that are not higher than third order in ρ, ρ':

$$\rho''(z) = \frac{1}{2}\frac{\Phi_0'}{\Phi_0}\rho' - \frac{1}{4}\frac{\Phi_0''}{\Phi_0}\rho + \mathcal{R}_3(\rho, \rho'; \Phi_0), \qquad (8.4b)$$

under the simplifying notation convention that Φ_0 is an abbreviation for $\Phi_{a0}(z)$, ρ for $\rho(z)$, and

$$\mathcal{R}_3(\rho, \rho'; \Phi_0) = \frac{1}{16}\left[\frac{1}{2}\frac{\Phi_0^{(4)}}{\Phi_0} - \left(\frac{\Phi_0''}{\Phi_0}\right)^2\right]\rho^3 + \frac{1}{8}\left[\frac{\Phi_0^{(3)}}{\Phi_0} - \frac{\Phi_0'\Phi_0''}{\Phi_0}\right)\rho^2\rho'$$
$$- \frac{1}{4}\frac{\Phi_0''}{\Phi_0}\rho\rho'^2 - \frac{1}{2}\frac{\Phi_0'}{\Phi_0}\rho'^3. \qquad (8.5)$$

The correction terms $\mathcal{R}_3(\rho, \rho'; \Phi_0)$ in (8.4) represent *third-order aberrations* from Gaussian optics (which results when they are neglected). Clearly each term in $\mathcal{R}_3(\rho, \rho'; \Phi_0)$ is proportional to a factor containing products of $\rho(z)$ and $\rho'(z)$ such that the sum of the powers of these two factors is exactly 3. That is the more precise meaning of the statement that each of these terms is of third order in ρ, ρ'. Note also that all terms in (8.4) are of dimension $(\textit{length})^{-1}$, but $\mathcal{R}_3$ is smaller in magnitude than the other two terms because it is of third order in the small parameters ρ, ρ'. If these parameters are sufficiently small, we obtain the Gaussian approximation which was discussed more fully in Chapter 5. The decision concerning "smallness" would have to be made by analyzing the magnitude of $\mathcal{R}_3$ as given in (8.5), each term of which contains a function of Φ_0 and its derivatives, as well as the third-order product of ρ, ρ' factors, that together yield terms of dimension $(\textit{length})^{-1}$ times a factor $(\rho/l)^3$. The smallness of ρ compared to the length l that has thus been defined artifically determines how important the correction term $\mathcal{R}_3$ is.

We obtain an approximate solution to (8.4) by means of *perturbation theory*. We shall perform a classical perturbation analysis in trying to solve that equation by assuming that $\rho(z)$ has two terms,

$$\rho(z) = \epsilon\xi(z) + \epsilon^3\delta\xi(z), \qquad (8.6)$$

where ϵ is a small parameter of the order of $\rho(z)/l$ (see above for the definition of l). Insertion of the definition (8.6) into (8.5) shows that $\mathcal{R}_3$ is of order ϵ^3, whereas the other terms in (8.5) are of order ϵ. Equating terms of $O(\epsilon)$ we thus observe that

$$\epsilon\xi(z) = \rho_p(z) , \tag{8.7}$$

where $\rho_p(z)$ must be a solution of the paraxial equation (6.1). If we then take all terms of $O(\epsilon^3)$ in (8.5), we obtain

$$\epsilon^3\left(\delta\xi'' + \frac{1}{2}\frac{\Phi_0''}{\Phi_0}\delta\xi' - \frac{1}{4}\frac{\Phi_0''}{\Phi_0}\delta\xi\right) = \mathcal{R}_3(\epsilon\rho_p, \epsilon\rho_p'; \Phi_0) . \tag{8.8}$$

At this stage it is useful to give $\epsilon^3\delta\xi(z)$ the name $\delta\rho(z)$, just as we have already named $\epsilon\xi(z) \equiv \rho_p(z)$, so that (8.6) can also be written as

$$\rho(z) = \rho_p(z) + \delta\rho(z) . \tag{8.9}$$

The advantage of this form is that we do not separate out explicitly the size parameter ϵ, whereas that is just what is useful in (8.6). At any rate, we can now rewrite (8.8) as

$$\delta\rho'' + \frac{1}{2}\frac{\Phi_0'}{\Phi_0}\delta\rho' - \frac{1}{4}\frac{\Phi_0''}{\Phi_0}\delta\rho = \mathcal{R}_3(\rho_p, \rho_p'; \Phi_0) , \tag{8.10}$$

and the error in not including more terms in (8.9) is essentially $O(\epsilon^3)$.

We show in Appendix II how to obtain a solution of (8.10) by the method of variation of parameters. The result is

$$\delta\rho(z) = \alpha_1(z)\rho_1(z) + \alpha_2(z)\rho_2(z) , \tag{8.11}$$

where $\rho_1(z)$ and $\rho_2(z)$ are two independent solutions of the paraxial equation (6.1) with initial conditions,

$$\rho_1(z) = 0 , \qquad \rho_2(0) = 0 , \qquad \rho_1'(0) = 0 , \qquad \rho_2'(0) = 1 , \tag{8.12}$$

and where the derivatives of $\alpha_1(z)$ and $\alpha_2(z)$ are given by

$$\begin{aligned} \alpha_1'(z) &= -\sqrt{\frac{\Phi_0(z)}{\Phi(0)}}\,\frac{\rho_2(z)}{\rho_1(0)}\,\mathcal{R}_3(\rho_p, \rho_p'; \Phi_0) , \\ \alpha_2'(z) &= +\sqrt{\frac{\Phi_0(z)}{\Phi(0)}}\,\frac{\rho_1(z)}{\rho_1(0)}\,\mathcal{R}_3(\rho_p, \rho_p'; \Phi_0) . \end{aligned} \tag{8.13}$$

With the solution (8.9) of (8.5), where $\delta\rho(z)$ is given by (8.11)–(8.13), the effect of aberrations upon the focal and image planes can be be analyzed.

8.1.1 Aberrations in the Focal Plane

Figure 8-1 deals with aberrations in the focal plane. The two reference trajectories $\rho_1(z)$ and $\rho_2(z)$ with initial conditions (8.12) are sketched as shown. We choose a $\rho(z)$ with the solution to the paraxial equation of motion,

$$\rho_p(z) \equiv \xi_0 \rho_1(z) , \tag{8.14}$$

where $\rho_1(z)$ is the first reference trajectory and ξ_0 is dimensionless. We then refer to Eqs. (8.13) to define the transverse coordinate

$$\rho(z) \equiv [\xi_0 + \alpha_1(z)]\rho_1(z) + \alpha_2(z)\rho_2(z) . \tag{8.15}$$

We shall require that $\alpha_1(0) = \alpha_2(0) = 0$ at the location defined as $z = 0$ in Fig. 8-1. This is easily satisfied outside the lensing region because $\delta\rho(z)$, defined in (8.11), is zero there. From the definitions (8.12) and (8.14) it follows that $\rho_1(z_f) = 0$ and $\rho_p(z_f) = 0$ at the location $z = z_f$ of the focal plane. Consequently,

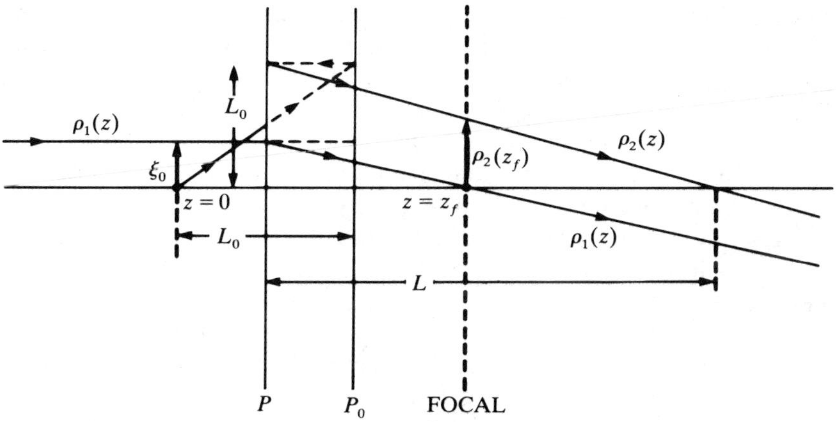

Fig. 8-1. The paraxial-optics sketch for treatment of spherical aberrations in the focal plane.

$$\rho(0) = \xi_0 \rho_1(0) , \qquad \rho(z_f) = \alpha_2(z_f)\rho_2(z_f) . \tag{8.16}$$

Apply (8.13) to the second of these equations to obtain,

$$\rho(z_f) = \rho_2(z_f) \int_0^{z_f} d\zeta \sqrt{\frac{\Phi_0(\zeta)}{\Phi(0)}} \frac{\rho_1(\zeta)}{\rho_1(0)} \mathcal{R}_3(\rho_p, \rho_p'; \Phi_0) . \tag{8.17}$$

Replace $\rho_p(\zeta)$ in the $\mathcal{R}_3$ factor by $\xi_0 \rho_1(\zeta)$ to change this factor into $\mathcal{R}_3(\rho_p, \rho_p'; \Phi_0) = \xi_0^3 \mathcal{R}_3(\rho_1, \rho_1'; \Phi_0)$. A glance at the geometry in Fig. 8-1 suffices to show that

$$\frac{\rho_2(z_f)}{L_0} = \frac{L-f}{L} = \left(\frac{L-f}{f}\right) \frac{f}{L} = M \frac{f}{L} , \tag{8.18}$$

where M is the magnification of the lens (see Eq. (6.9)). With these two substitutions in (8.17), we find

$$\rho(z_f) = C_{fa} \xi_0^3 ,$$

$$C_{fa} = ML_0 \left\{ \frac{f}{L} \int_0^{z_f} d\zeta \sqrt{\frac{\Phi_0(\zeta)}{\Phi(0)}} \frac{\rho_1(\zeta)}{\rho_1(0)} \mathcal{R}_3(\rho_1, \rho_1'; \Phi_0) \right\} .$$

Unlike the solution to the paraxial equation, the inclusion of third-order terms introduces a nonzero value for $\rho(z)$ at the focal point, and $\rho(z_f)$ is known as the *third-order ray aberration* in the focal plane. The quantity C_{fa} with a dimension of (*length*) is the *coefficient of third-order aberrations* in the focal plane. The factor ML_0 is not always included in the aberration coefficient; various texts follow differing conventions. Calculation of C_{fa} requires knowledge of the axial profile of the potential $\Phi_0(z)$, and of the two principal trajectories $\rho_1(z)$ and $\rho_2(z)$. The quality of a lens is determined partly by favorably low values of $|C_{fa}|$.

In general, $C_{fa} < 0$. To make this plausible, consider the integrand in (8.19). Figure 8-1 shows that the ratio $\rho_1(\zeta)/\rho_1(0)$ is not less than zero in $0 < z < z_f$. The quantity $\mathcal{R}_3$ is determined chiefly by the first term in (8.5) whenever it is true that $|\rho'| \ll |\rho/l|$, where l is a length determined by the derivatives of Φ_0. This first term tends to be negative because the contribution of the term with a fourth derivative of Φ_0 is usually much smaller than the other one. Consequently, $\mathcal{R}_3 < 0$, and thus $C_{fa} < 0$ and $\rho(z_f) < 0$. The third-order aberrations cause an *overfocussing* of trajectories through the lens, i.e., the true focus occurs closer to the object than the paraxial-ray focus. However, this is only a plausibility argument that holds for many common types of axisymmetric electrostatic lenses, and the reader is referred to a discussion of Scherzer's theorem (Grivet 1972) for a more general proof that all axisymmetric electrostatic lenses give rise to a negative aberration coefficient.

8.1.2 Aberrations in the Image Plane

The procedure for trajectories that image a point paraxially into another point is somewhat similar. In this case we follow a ray that behaves paraxially as a small multiple of $\rho_2(z)$, i.e., we choose $\rho(z)$ as a sum of the parts defined in (8.9) and (8.11) with

$$\rho_p(z) = (\tan\theta_0)\rho_2(z) \approx \theta_0\rho_2(z)\,, \tag{8.20}$$

so that we obtain

$$\rho(z) = \alpha_1(z)\rho_1(z) + [\theta_0 + \alpha_2(z)]\rho_2(z)\,. \tag{8.21}$$

At the location of the image plane $z = z_i$, we have $\rho_2(z_i) = 0$, which yields via (8.13),

$$\rho(z_i) = \alpha_1(z_i)\rho_1(z_i) = -\rho_1(z_i)\int_0^{z_i} d\zeta\,\sqrt{\frac{\Phi_0(\zeta z)}{\Phi(0)}}\,\frac{\rho_2(\zeta)}{\rho_1(0)}\,\mathcal{R}_3(\rho_p, \rho_p'; \Phi_0)\,. \tag{8.22}$$

By virtue of (8.20) we have $\mathcal{R}_3(\rho_p, \rho_p'; \Phi_0) = \theta_0^3\mathcal{R}_3(\rho_2, \rho_2'; \Phi_0)$. A glance at Fig. 6-1 shows that $\rho_1(z_i) = -M\rho_1(0)$, where $M = (L - f)/f = f_0/(L_0 - f_0)$ is the magnification of the lens as given in (6.9). As a consequence, (8.22) can be rewritten as

$$\rho(z_i) = C_{ia}\theta_0^3\,,$$

$$C_{ia} = M\int_0^{z_i} d\zeta\,\sqrt{\frac{\Phi_0(\zeta z)}{\Phi(0)}}\,\rho_2(\zeta)\mathcal{R}_3(\rho_2, \rho_2'; \Phi_0)\,. \tag{8.23}$$

Here too, the aberration coefficient C_{ia} has the dimension of (*length*). As before, some authors exclude the magnification factor M from a definition of the aberration coefficient. For reasons similar to those given in connection with (8.19), we find that C_{ia} is negative because $\rho_2(\zeta) > 0$ in the lensing region. The rays overfocus more strongly as angle θ_0 increases. Here too, C_{ia} is always negative for axisymmetric focussing lenses.

Let us consider a transition from the situation of Fig. 8-2 to Fig. 8-1 by moving the object point from $z = 0$ to $z = -\infty$. The image point will move

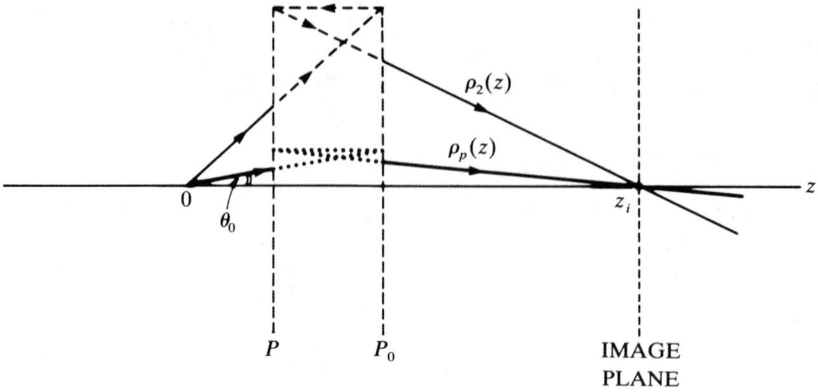

Fig. 8-2. The paraxial-optics sketch for treatment of spherical aberrations in the image plane.

from $z = z_i$ to $z = z_f$. Will the formula (8.23) go over into (8.19)? Clearly, we must replace θ_0 by ξ_0/L_0, even though $L_0 \to \infty$. However, $\rho_2(\zeta) \approx L_0$ in the lensing region, and a factor $\xi_0/L_0 = f/L$. This will suffice as $\rho_1(0) = 1$ can be placed in the denominator. At any rate, (8.19) and (8.23) represent different manifestations of one effect: stronger deflection towards the axis than would be the case when calculated by paraxial optics. This effect is known more precisely as *spherical aberration* (see Section 8.2).

8.2 GENERAL ABERRATION THEORY FOR AXISYMMETRIC SYSTEMS

In this section we will again approach the subject of deviations from paraxial optics, but from a more general point of view. To do so, we introduce the generalized transverse position coordinates $\vec{q}(z)$, and momentum coordinates $\vec{p}(z)$, as we did in Section 3.2. All four of these coordinates are functions, in principle, of the reference coordinate z (so named because we shall restrict ourselves shortly to axisymmetric systems). Consider the trajectory sketched in Fig. 8-3: the initial transverse coordinates are $\vec{q} = \vec{q}(0)$, $\vec{p} = \vec{p}(0)$, and the final ones are $\vec{Q} = \vec{q}(L)$, $\vec{P} = \vec{p}(L)$. We may consider $\vec{Q}(\vec{q}, \vec{p})$ and $\vec{P}(\vec{q}, \vec{p})$ to be functions of the initial coordinates, as they obviously do depend upon them. The $\vec{q}$, $\vec{p}$ notation is often used in the classical Hamilton-Jacobi formulation of mechanics (Corbon and Stehle 1960), but we restrict its use here to only the transverse coordinates. Classical physics states that $\vec{Q}$ and $\vec{P}$ are

Fig. 8-3. Nomenclature for initial and final coordinates of an electron trajectory.

determined entirely by the initial values $\vec{q}$ and $\vec{p}$ through the equations of motion, hence the above stated functional dependence in the notation. It goes without saying that the dependence upon electromagnetic forces is implicit; that manifests itself explicitly in the equations of motion. For the sake of convenience, we will work in Cartesian coordinates, so that first-order optics predicts for any small displacement of the four initial coordinates a corresponding displacement of the four final ones according to the matrix

$$\begin{bmatrix} \delta Q_x \\ \delta P_x \\ \delta Q_y \\ \delta P_y \end{bmatrix} = \begin{bmatrix} \dfrac{\partial Q_x}{\partial q_x} & \dfrac{\partial Q_x}{\partial p_x} & \dfrac{\partial Q_x}{\partial q_y} & \dfrac{\partial Q_x}{\partial p_y} \\ \dfrac{\partial P_x}{\partial q_x} & \dfrac{\partial P_x}{\partial p_x} & \dfrac{\partial P_x}{\partial q_y} & \dfrac{\partial P_x}{\partial p_y} \\ \dfrac{\partial Q_y}{\partial q_x} & \dfrac{\partial Q_y}{\partial p_x} & \dfrac{\partial Q_y}{\partial q_y} & \dfrac{\partial Q_y}{\partial p_y} \\ \dfrac{\partial P_y}{\partial q_x} & \dfrac{\partial P_y}{\partial p_x} & \dfrac{\partial P_y}{\partial q_y} & \dfrac{\partial P_y}{\partial p_y} \end{bmatrix} \begin{bmatrix} \delta q_x \\ \delta p_x \\ \delta q_y \\ \delta p_y \end{bmatrix} \tag{8.24}$$

As an example consider the focal plane for principal electrons as the final location of trajectories. At the initial plane in front of the lensing region, we have $\delta p_x = \delta p_y = 0$ for principal electrons. Then δq_x and δq_y represent infinitesimal displacements from one electron trajectory to another at $z = 0$. We know that Gaussian optics will then predict that $\delta Q_x = \delta Q_y = 0$ in the focal plane, so that the only part of (8.24) that remains is

$$\begin{bmatrix} \delta P_x \\ \delta P_y \end{bmatrix} = \begin{bmatrix} \dfrac{\partial P_x}{\partial q_x} & \dfrac{\partial P_x}{\partial q_y} \\ \dfrac{\partial P_y}{\partial q_x} & \dfrac{\partial P_y}{\partial q_y} \end{bmatrix} \begin{bmatrix} \delta q_x \\ \delta q_y \end{bmatrix}. \tag{8.25}$$

In Fig. 8-4, we have sketched two principal-electron trajectories in a

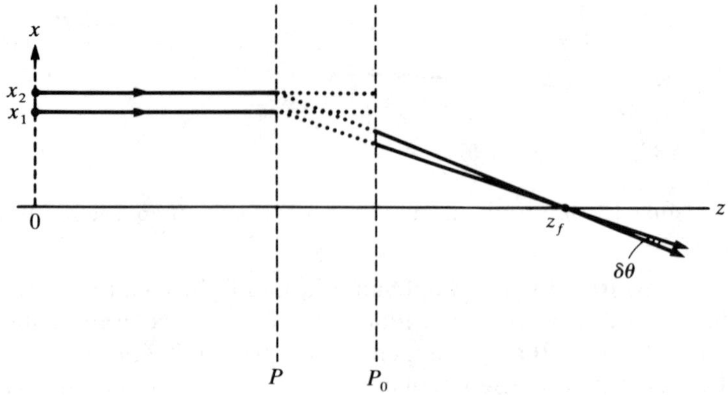

Fig. 8-4. Two closely spaced principal electrons passing through a thin lens.

$y = \text{constant}$ plane such that $\delta x = x_2 - x_1$ is very small at $z = 0$. To first order, we have $\delta\theta = \delta x \cos\theta/f$. The momentum at $z = z_f$ is $P_x = P_z \tan\theta$, and therefore $\delta P_x = (P_z/\cos^2\theta)\,\delta\theta = P_z \delta x/(f\cos\theta) = P\delta x/f$ (if P is the total momentum). For the y motion something similar holds, but here we obtain

$$\begin{bmatrix} \delta P_x \\ \delta P_y \end{bmatrix} = \begin{bmatrix} \dfrac{P}{f} & 0 \\ 0 & 0 \end{bmatrix} \begin{bmatrix} \delta q_x \\ \delta q_y \end{bmatrix}. \tag{8.26}$$

Here, we have identified $\delta q_x \equiv \delta x$ and $\delta q_y \equiv \delta y$. Thus, (8.26) is a fairly trivial example that identifies the four matrix elements in (8.25) explicitly.

If aberrations are to be considered, then it will be necessary to expand $\delta\vec{Q}$ and $\delta\vec{P}$ into higher powers of the components of $\delta\vec{q}$ and $\delta\vec{p}$. In the case of axial symmetry, we expect $\vec{Q}$ and $\vec{P}$ to be functions only of the magnitudes $q \equiv |\vec{q}|$ and $p \equiv |\vec{p}|$ and of the angle ϕ with $\cos\phi = \vec{q}\cdot\vec{p}/qp$. If we agree to look only at meridional-electron trajectories (i.e., the total momentum vector lies in a plane with the z axis) then we have $\phi = 0$ and the final values of the coordinates $\vec{Q}, \vec{P}$ are functions only of the magnitudes q, p of the initial coordinates. Specifically, we obtain through third order,

$$\begin{aligned} \delta Q &\equiv Q(q + \delta q, p + \delta p) \\ &= \frac{\partial Q}{\partial q}\,\delta q + \frac{\partial P}{\partial p}\,\delta p + \frac{1}{2}\,\frac{\partial^2 Q}{\partial q^2}\,(\delta q)^2 + \frac{1}{2}\,\frac{\partial^2 Q}{\partial p^2}\,(\delta p)^2 \end{aligned}$$

$$+\frac{\partial Q^2}{\partial q\,\partial p}\,\delta q\,\delta p+\frac{1}{6}\,\frac{\partial Q^3}{\partial q^3}\,(\delta q)^3+\frac{1}{2}\,\frac{\partial Q^3}{\partial q^2\,\partial p}\,(\delta q)^2\,\delta p$$

$$+\frac{1}{2}\,\frac{\partial Q^3}{\partial q\,\partial p^2}\,\delta q(\delta p)^2+\frac{1}{6}\,\frac{\partial Q^3}{\partial p^3}\,(\delta p)^3+\cdots. \tag{8.27}$$

Consider an opposing excursion $\delta Q_1 = Q(-q-\delta q, -p-\delta p) - Q(-q, -p)$. The axial symmetry forces us to conclude that $\delta Q_1 = -\delta Q$. If $\delta q_1 = -\delta q$ and $\delta p = -\delta p$, then it follows that $\partial^{(m+n)}Q_1/\partial q_1^m\,\partial p_1^n = \partial^{(m+n)}Q/\partial q^m\,\partial p^n$ for all the derivatives in (8.27). As a consequence, all terms in (8.27) for which $m+n$ is even must be zero if δQ_1 is to be equal to $-\delta Q$. The quadratic terms in (8.27) must then be zero, and we can rewrite that equation as

$$\delta Q = k_{11}\,\delta q + k_{12}\,\delta p + k_{31}(\delta q)^3 + k_{32}(\delta q)^2\,\delta p + k_{33}\,\delta q(\delta p)^2$$
$$+ k_{34}(\delta p)^3\,, \tag{8.28}$$

where the coefficients $k_{11}, \ldots, k_{34}$ are given explicitly by the first-order and third-order terms in (8.27). All this, however, holds only for axial symmetry. In the more general case, i.e., trajectories that are not meridional, we find that the angle ϕ between $\delta\vec{q}$ and $\delta\vec{p}$ plays a role in the expressions, and that (8.28) must be replaced by a more general expression for dQ_x and δQ_y separately. The axial symmetry still dictates that quadratic terms in the components δq_i and δp_i are absent by means of similar arguments. In fact, the expression for δQ_x can be generated from (8.27) upon replacing the factors δq and δp by the linear combinations,

$$\begin{aligned}\delta q &\rightarrow \alpha_{11}\,\delta q_x + \alpha_{12}\,\delta q_y + \alpha_{13}\,\delta p_x + \alpha_{14}\,\delta p_y\,,\\ \delta p &\rightarrow \alpha_{21}\,\delta q_x + \alpha_{22}\,\delta q_y + \alpha_{23}\,\delta p_x + \alpha_{24}\,\delta p_y\,,\end{aligned} \tag{8.29}$$

and by subsequently collecting all terms $\delta q_x^k\,\delta q_y^l\,\delta p_x^m\,\delta p_y^n$ which have the same values of k, l, m, n in the exponent together into one term with only one coefficient. The resulting *third-order* contribution to δQ_x is

$$\begin{aligned}\delta Q_x^{(3)} = {}&(a_1\,\delta q_x^3 + a_2\,\delta q_x^2\,\delta q_y + a_3\,\delta q_x\,\delta q_y^2 + a_4\,\delta q_y^3)\\ &+ (b_1\,\delta q_x^2\,\delta p_x + b_2\,\delta q_x\,\delta q_y\,\delta p_x + b_3\,\delta q_y^2\,\delta p_x + b_4\,\delta q_x^2\,\delta p_y\\ &\quad + b_5\,\delta q_x\,\delta q_y\,\delta p_y + b_6\,\delta q_y^2\,\delta p_y)\\ &+ (c_1\,\delta p_x^2\,\delta q_x + c_2\,\delta p_x\,\delta p_y\,\delta q_x + c_3\,\delta p_y^2\,\delta q_x + c_4\,\delta p_x^2\,\delta q_y\\ &\quad + c_5\,\delta p_x\,\delta p_y\,\delta q_y + c_6\,\delta p_y^2\,\delta q_y)\\ &+ (d_1\,\delta p_x^3 + d_2\,\delta p_x^2\,\delta p_y + d_3\,\delta p_x\,\delta p_y^2 + d_4\,\delta p_y^3)\,.\end{aligned} \tag{8.30}$$

Note that there are no duplications, although the procedure described above (8.30) would lead to many duplicate terms before the gathering of all the ones with identical exponents behind a single new coefficient.

The next point to be noted is that axial symmetry forces $\delta Q_x^{(3)}$ to be only a function of q, p, and ϕ, as explained above in connection with (8.27). As a result of this restriction, some of the terms in (8.30) should not appear by themselves, or, to put it equivalently but differently, some of the coefficients in (8.30) are not independent. For example, consider the term $a_1\,\delta q_x^3 = \alpha_1\,\delta q^3(\cos\phi_q)^3$, where $\cos\phi_q \equiv \hat{x}\cdot\delta\vec{q}/\delta q$ (if $\hat{x}$ is a unit vector along the x axis). This term should combine ultimately with others in $\delta Q_x^{(3)}$ and $\delta Q_y^{(3)}$ such that $\delta\vec{Q}^{(3)}$ is a function of angle $\phi = \phi_q - \phi_p$ rather than of ϕ_q or ϕ_p (if $\cos\phi_p \equiv \hat{x}\cdot\delta\vec{p}/\delta p$).

The various interdependencies of the coefficients of $\delta Q_x^{(3)}$ and $\delta Q_y^{(3)}$ are made explicitly visible most economically by use of complex notation. The vector $\delta\vec{q} = \hat{x}\,\delta q_x + \hat{y}\,\delta q_y$ can be replaced by a *phasor* $\delta q_c = \delta q\exp(i\phi_q)$ so that $\mathrm{Re}[\delta q_c]$ corresponds to δq_x and $\mathrm{Im}[\delta q_c]$ to δq_y. The phase angle ϕ_q is still the angle between $\delta\vec{q}$ and the unit vector $\hat{x}$, and the magnitude is still $\delta q = \sqrt{\delta q_x^2 + \delta q_y^2}$. The geometry of complex phasors preserves the geometry of two-dimensional vectors completely. Let us therefore define

$$\begin{aligned}
\delta Q_c &\equiv \delta Q_x^{(3)} + i\,\delta Q_y^{(3)}\,, \\
\delta q_c &\equiv \delta q_x + i\,\delta q_y = \delta q\exp(i\phi_q)\,, \\
\delta p_c &\equiv \delta p_x + i\,\delta p_y = \delta p\exp(i\phi_p)\,.
\end{aligned} \tag{8.31}$$

The dot product $\delta\vec{q}\cdot\delta\vec{p}$ is represented by $\mathrm{Re}[\delta q_c^*\,\delta p_c]$, as is easily verified from (8.31). The cross product $|\delta\vec{q}\times\delta\vec{p}|$ is represented by $\mathrm{Im}[\delta q_c^*\,\delta p_c]$. Note that both of these quantities involving vectors are functions of the angles ϕ_q and ϕ_p only through their difference $\phi_q - \phi_p \equiv \phi$, and consequently $\delta q_c^*\,\delta p_c$ (or its complex conjugate) is also a function of ϕ rather than of ϕ_q and ϕ_p separately. A rotation of the electron-trajectory displacement vectors $\delta\vec{q}$ and $\delta\vec{p}$ simultaneously over angle ψ will produce new values of $\delta q_c^*\,\delta p_c$ that depend upon ψ. Incidentally, we do expect $\delta\vec{Q}$ and $\delta\vec{P}$ to rotate by the same angle ψ under these circumstances. For that reason, δQ_c will rotate into $\delta Q_c\exp(i\psi)$. This fact needs to be incorporated into our third-order expressions, of which only $\delta Q_x^{(3)}$ is given explicitly in (8.30). When it is, a third-order form in terms of the above complex notation that is in agreement with the rotation property is

$$\begin{aligned}
\delta Q_c = (\beta_1\,\delta q_c + \beta_2\,\delta p_c)(&\beta_3\,\delta q_c\,\delta q_c^* + \beta_4\,\delta q_c\,\delta p_c^* \\
&+ \beta_5\,\delta q_c^*\,\delta p_c + \beta_6\,\delta p_c\,\delta p_c^*\,.
\end{aligned} \tag{8.32}$$

Note that the quadratic factor is the most general one that is independent of the rotation angle ψ, and that the linear factors must have the same dependence upon ψ that δq_c and δp_c have. Thus (8.32 must generate the desired dependence of $\delta \vec{Q}$ upon $\delta \vec{q}$ and $\delta \vec{p}$ in the third-order terms. Further reduction of (8.32) by multiplying out the various terms and collecting duplicates behind one single new coefficient yields

$$\delta Q_c = a|\delta p_c|^2\, \delta p_c + b\, \delta q_c^2\, \delta p_c^* + c|\delta q_c|^2\, \delta p_c + d\, \delta p_c^2\, \delta q_c^* \\ + e|\delta p_c|^2\, \delta q_c + f|\delta q|^2\, \delta q_c\,. \tag{8.33}$$

There are *twelve* terms with independent coefficients in (8.33) because each of the six coefficients $a, b, \ldots, f$ is complex in principle. This immediately reduces to *six* apparently independent coefficients in *electrostatics* because meridional electrons travel in one meridional plane with $\phi_q = \phi_p = \text{constant}$. We can see this, because it implies that all terms on both sides of (8.33) must be real after a factor $\exp(i\phi_q)$ has been divided out. That makes a, b, c, d, e, and f real, and if we then take the real part of (8.33), we obtain after translation back into vector notation,

$$\delta Q_x^{(3)} = A\, \delta p^2\, \delta p_x + B[(\delta q_x^2 - \delta q_y^2)\, \delta p_x + 2\, \delta q_x\, \delta q_y\, \delta p_y] \\ + C\, \delta q^2\, \delta p_x + D[(\delta p_x^2 - \delta p_y^2)\, \delta q_x + 2\, \delta p_x\, \delta p_y\, \delta q_y] \\ + E\, \delta p^2\, \delta q_x + F\, \delta q^2\, \delta q_x\,.$$

A similar expression for $\delta Q_y^{(3)}$ can be obtained by interchanging all subindices $x \leftrightarrow y$. The six new coefficients A, B, C, D, E, and F are real, but they are not all independent. We shall see that $E = 2D$. Contrary to what is implied implicitly in several textbooks (Grivet 1972, Born and Wolf 1965) by referring this property to a theorem by Malus and Dupin, this is not at all obvious, and we devote the remainder of this subsection to this.

8.2.1 Proof that $E = 2D$ in Eq. (8.34) and Its Consequence

Let us consider the very general geometrical optics trajectory (2.6) which applies to the trajectory of an electron when $n(s) \propto \sqrt{\Phi_a(s)}$, as indicated in (2.29). From (2.5), we then obtain

$$n(s)\,\frac{d\vec{r}}{ds} = \vec{\nabla} S\,, \tag{8.35}$$

where S is the *eikonal* function introduced in (2.3). Upon integration of (8.35) we obtain

$$\vec{r}(s) = \vec{r}(0) + \int_0^s ds_1 \, \frac{1}{n(s_1)} \, (\vec{\nabla} S)_{s=s_1} \, . \tag{8.36}$$

If we define this particular trajectory to be the reference trajectory, and if $\vec{q}(s)$ is a coordinate on an orthogonal surface that is normal to the reference trajectory then of course $\vec{q}(s) = 0$ for the reference trajectory. Nevertheless, we can write

$$\vec{Q} = \vec{q} + \int_0^s ds_1 \, \frac{1}{n(s_1)} \, (\vec{\nabla}_q S)_{s=s_1} \, , \tag{8.37}$$

even though $\vec{Q} \equiv \vec{q}(s)$ and $\vec{q} \equiv \vec{q}(0)$ are essentially zero. If (see Fig. 8-5) we define a neighboring trajectory that starts at $\vec{q} + \delta\vec{q}$ with momentum $\vec{p} + \delta\vec{p}$, then the difference equation obtained from (8.37) is

$$\delta\vec{Q} = \delta\vec{q} + \int_0^s ds_1 \, \frac{1}{n(s_1)} \, (\vec{\nabla}_q \delta S)_{s=s_1} \, . \tag{8.38}$$

Now we make use of the fact that $S(\vec{q}, \vec{p})$ is a function of the *initial* coordinates to describe an infinitesimal variation in S as

$$\delta S = \frac{\partial S}{\partial q_x} \, \delta q_x + \frac{\partial S}{\partial q_y} \, \delta q_y + \frac{\partial S}{\partial p_x} \, \delta p_x + \frac{\partial S}{\partial p_y} \, \delta p_y \, . \tag{8.39}$$

It is now not difficult to verify from (8.38) and (8.39) that

$$\frac{\partial \delta Q_x}{\partial \delta p_y} = \frac{\partial \delta Q_y}{\partial \delta p_x} \, , \tag{8.40}$$

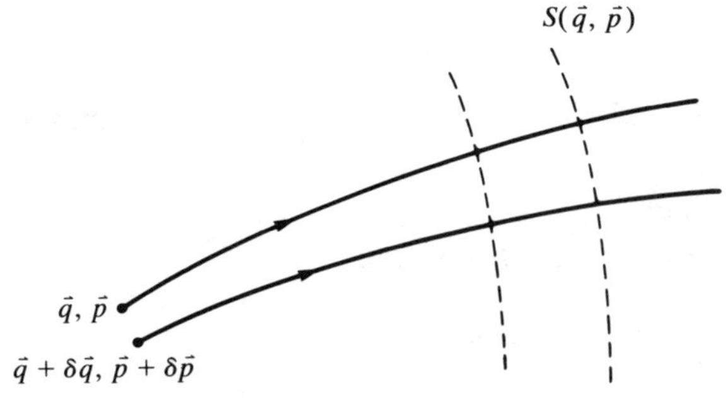

Fig. 8-5. Nomenclature for two neighboring trajectories, and the eikonal function S.

because the definition of the reference trajectory implies that $\vec{\nabla}_q = \vec{\nabla}_{\delta q}$ in (8.38). The next step is to apply (8.40) to the third-order terms of (8.34) to obtain

$$\frac{\partial \delta Q_x^{(3)}}{\partial \delta p_y} = 2A\,\delta p_x\,\delta p_y + 2B\,\delta q_x\,\delta q_y + 2D(-\delta q_x\,\delta p_y + \delta q_y\,\delta p_x) + 2E\,\delta q_x\,\delta p_y\,,$$

$$\frac{\partial \delta Q_y^{(3)}}{\partial \delta p_x} = 2A\,\delta p_x\,\delta p_y + 2B\,\delta q_x\,\delta q_y + 2D(-\delta q_y\,\delta p_x + \delta q_x\,\delta p_y) + 2E\,\delta q_y\,\delta p_x\,. \tag{8.41}$$

Subtract these two expressions to obtain

$$0 = (4D - 2E)(\delta q_y\,\delta p_x - \delta q_x\,\delta p_y)\,, \tag{8.42}$$

and it follows immediately that the right-hand side is zero only when $E = 2D$.

After some rearrangement of terms, (8.34) and the similar expression for $\delta Q_y^{(3)}$ combine into

$$\delta\vec{Q}^{(3)} = A\,\delta p^2\,\delta\vec{p} + [\delta p^2\,\delta\vec{p} + 2(\delta\vec{q}\cdot\delta\vec{p})\,\delta\vec{p}] + (C - B)\,\delta q^2\,\delta\vec{p} + 2B(\delta\vec{q}\cdot\delta\vec{p})\,\delta\vec{q} + F\,\delta q^2\,\delta\vec{q}\,. \tag{8.43}$$

The final result shows that there are only *five* independent aberration coefficients for axisymmetric electrostatic lenses.

8.2.2 Spherical Aberrations

Consider a point object on the axis so that $\delta\vec{q} = 0$, and let it emit a cone of electron trajectories with transverse momentum $\delta\vec{p} = \delta p \exp(i\phi_p)$, written in complex notation, where δp is constant and ϕ_p is distributed uniformly between 0 and 2π. The only term remaining in (8.43) is the first, and the third-order aberrations at $z = L$ then reduce to $\delta\vec{Q}^{(3)} = A\,\delta p^2\,\delta\vec{p}$, i.e., to a circle with radius $A\,\delta p^3$. If we allow the magnitude of the momentum δp also to vary between 0 and p_0, then a circle with radius Ap_0^3 will be filled at $z = L$, as shown in Fig. 8-6. These *spherical aberrations* have been discussed more extensively in Section 8.1.2, and A represents the coefficient of spherical aberrations when these are expressed in terms of the transverse momentum.

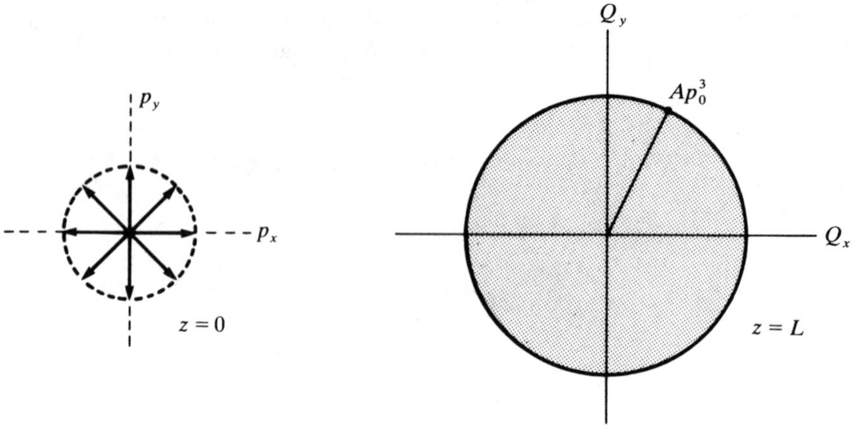

Fig. 8-6. *Spherical aberration* of a point focus.

8.2.3 Distortion

Consider a linear object of length δq at right angles to the z axis at $z = 0$. Principal electrons emitted from this object will image this length as $F\,\delta q^3$ at $z = L$, and the object will have remained in the same plane. The square of principal trajectories shown in Fig. 8-7 at $z = 0$ will image as shown in the two other sketches (one for $F > 0$ and one for $F < 0$). The corners of the square at $z = 0$ will image further away from the axis when $F > 0$ than the centers of the side, and less far when $F < 0$, to give a typical "pincushion" and "barrel" shape. This is known as *distortion*, and F is the coefficient of distortion.

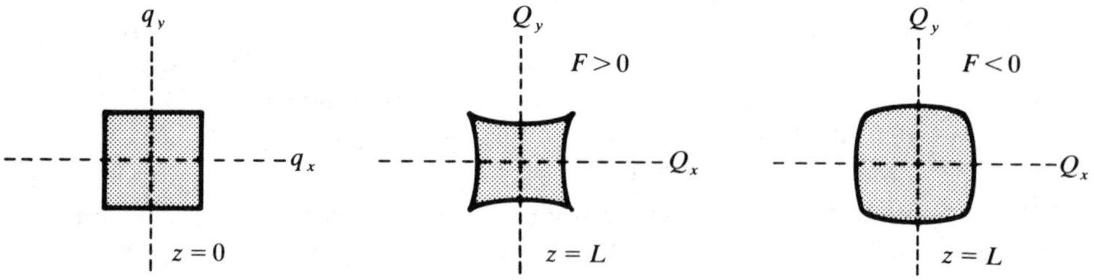

Fig. 8-7. Pincushion (2nd figure) and barrel (3rd figure) *distortion* of what would otherwise have been a square.

8.2.4 Coma

This effect ocurs when the term $D[\delta p^2\, \delta\vec{q} + 2(\delta\vec{q} \cdot \delta\vec{p})\, \delta\vec{p}]$ is dominant. That might be the case when all δq are sufficiently small so that only the first-order terms in δq are important in (8.34), but also δp is sufficiently small so that the spherical-aberration term is small. At any rate, consider the effect of this term by itself. To do so in more detail, write it as

$$\begin{aligned} \delta Q_x^{(3)} &= D\, \delta q\, \delta p^2[\cos\phi_q + 2\cos(\phi_q - \phi_p)\cos\phi_p]\,, \\ \delta Q_y^{(3)} &= D\, \delta q\, \delta p^2[\sin\phi_q + 2\cos(\phi_q - \phi_p)\sin\phi_p]\,. \end{aligned} \tag{8.44}$$

Consider electrons emitted in a cone from a point on the x axis ($\phi_q = 0$). Then (8.44) becomes

$$\begin{aligned} \delta Q_x^{(3)} &= D\, \delta q\, \delta p^2(2 + \cos 2\phi_p)\,, \\ \delta Q_y^{(3)} &= D\, \delta q\, \delta p^2 \sin 2\phi_p\,. \end{aligned} \tag{8.45}$$

Figure 8-8 shows the circle at $z = L$ that results from the above-described cone of trajectories emitted at $\delta\vec{q} = \hat{x}\, \delta q$ with $\delta\vec{p} = \delta p \exp(i\phi_p)$ (in complex notation), given that $0 < \phi_p < 2\pi$. Expressions (8.45) show that each point on this circle is intersected *twice*, once by a trajectory defined by a value of ϕ_p between 0 and π at $z = 0$, and again by the trajectory launched at $\phi_p + \pi$. These expressions also tell us how a uniformly filled cone of electrons, emitted from points with $\delta q < q_0$ on the x axis and with transverse momentum $\delta p < p_0$, will image at $z = L$. This filled cone

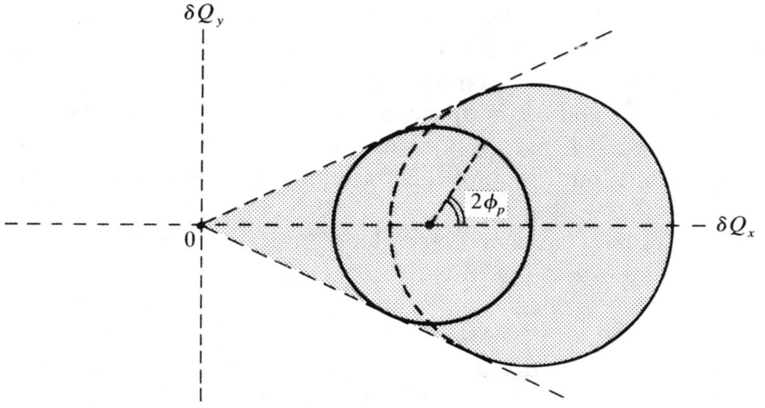

Fig. 8-8. Illustration of *coma* at the image plane.

produces the (larger) shaded image in Fig. 8-8; it is produced by continuously moving a filled circle with center on the δQ_x axis to the right. The circle at all locations just fits inside the wedge of two straight lines, each with angle of 30° with the δQ_x axis; the movement to the right continues until the radius of the circle is $Dq_0 p_0^2$. The shape of this wedge-circle figure is known as *coma* and D is its coefficient.

8.2.5 Curvature and Astigmatism

The aberrations produced by small transverse components of velocity in trajectories that do not necessarily originate close to the axis are governed by

$$\delta \vec{Q}^{(3)} = (C - B)\, \delta q^2\, \delta \vec{p} + 2B(\delta \vec{q} \cdot \delta \vec{p})\, \delta \vec{q}\ . \tag{8.46}$$

Note that these are the terms quadratic in δq and linear in δp. In terms of the two Cartesian components, we have

$$\begin{aligned} \delta Q_x^{(3)} &= \delta q^2\, \delta p[C \cos \phi_p + B \cos(2\phi_q - \phi_p)]\ , \\ \delta Q_y^{(3)} &= \delta q^2\, \delta p[C \sin \phi_p + B \sin(2\phi_q - \phi_p)]\ . \end{aligned} \tag{8.47}$$

The length of this vector is given by

$$\delta Q^{(3)} = \delta q^2\, \delta p \sqrt{C^2 + B^2 + 2BC \cos[2(\phi_q - \phi_p)]}\ . \tag{8.48}$$

Equations (8.47) and (8.48) represent an ellipse in the Q_x, Q_y plane as ϕ_p is varied from 0 to 2π. That situation can be pictured by considering trajectories that have a fixed vector $\delta \vec{q}$, fixed value of magnitude $|\delta \vec{p}|$, but variable ϕ_p. This represents a circular cone of trajectories emanating from a point defined by fixed $\delta \vec{q}$ with transverse momentum components $\delta p_x = \delta p \cos \phi_p$, and $\delta p_y = \delta p \sin \phi_p$ (where δp is fixed but $0 < \phi_p < 2\pi$). The ellipse at $z = L$ has a major axis $|C + B|$ for $\phi_q = \phi_p$, and a minor axis $|C - B|$ if C and B are of the same polarity. More specifically, points $P_1, \ldots, P_4$ in the object plane at $z = 0$ in Fig. 8-9a image into points $P_1', \ldots, P_4'$ in the Gaussian image plane around the Gaussian image point at $z = L$, as shown in Fig. 8-9b.

Note that the tilt of the major axis is also at angle ϕ_q. If we continue the trajectories $P_1 \rightarrow P_1'$ and $P_2 \rightarrow P_2'$ beyond the image plane, then these will intersect at some distance δL_{12} beyond it. Likewise, the trajectories $P_3 \rightarrow P_3'$ and $P_4 \rightarrow P_4'$, when extended, will intersect at distance δL_{34} (see Fig. 8-10). The best image is a circle somewhere between δL_{12} and δL_{34}

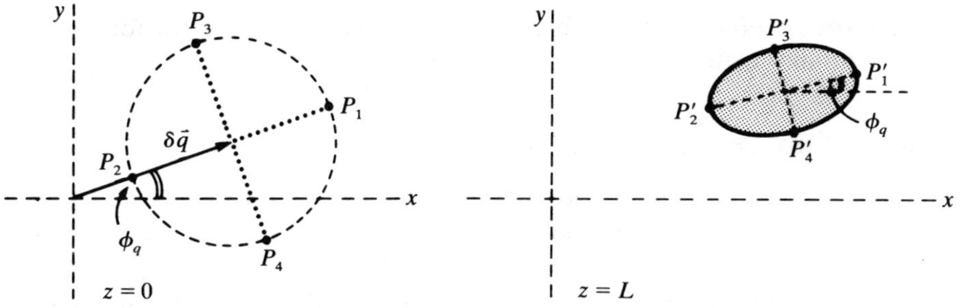

Fig. 8-9. *Field curvature* of an initially round shape.

beyond the Gaussian image plane. It represents a compromise in that neither δQ_x nor δQ_y are zero but each is smaller than its linear image at δL_{12} or δL_{34}. The position of this *circle of least confusion* will be a function of the size of δq because focussing effects are stronger for larger δq, which will make this circle lie closer to the image plane. The axial symmetry thus produces a *curved asymmetric surface* as locus of all circles of least confusion imaged by $\delta\vec{p}$ cones from $z = 0$ for $0 < \delta q < \delta q_{\max}$ where $\delta q_{\max}$ is a maximum aperture radius of, say, a circular object aperture. This surface will be *concave* towards the *object* aperture and its precise shape is not easily specified.

Equation (8.47) would yield a circle for the figure on which $P_1', \ldots, P_4'$ of Fig. 8-9b lie when coefficient $B = 0$. The circle of least confusion might collapse to a point, but that point would still lie on a curved asymmetric surface. Therefore, we say that coefficient C represents *field curvature*. On the other hand, coefficient B is responsible for the separation of foci in and perpendicular to a meridional plane, as shown in Fig. 8-10.

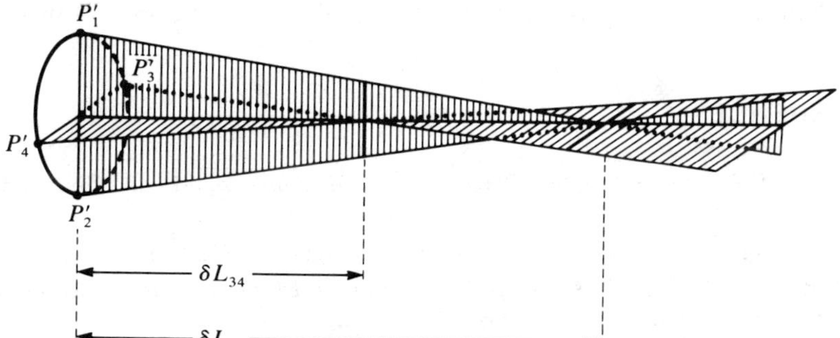

Fig. 8-10. *Astigmatic* imaging of an initially round shape. Note circle of least confusion between vertical and horizontal images.

Therefore, coefficient B represents *astigmatism*, i.e., different foci in the two mutually orthogonal planes of interest.

8.2.6 Chromatic and Other Aberrations

In wave optics, *chromatic aberration* refers to the effect caused by the wavelength dependence of the refractive index. Waves of different frequencies (or wavelengths) will image or focus in different planes. In electron optics, the accelerating potential may differ from one electron to the next because the initial velocities differ. Hence we will have $\Phi_a + \delta\Phi$ instead of Φ_a in (8.1) and (8.4) is replaced by

$$\rho''(z) = \frac{1}{2}\,\frac{\Phi_0'}{\Phi_0}\,\rho' - \frac{1}{4}\,\frac{\Phi_0''}{\Phi_0}\,\rho + \mathcal{R}_{ca}(\rho, \rho'; \Phi_0)\,, \tag{8.49}$$

with the aberration term

$$\mathcal{R}_{ca}(\rho, \rho'; \Phi_0) \equiv \frac{\delta\Phi}{\Phi_0}\left(\frac{1}{2}\,\frac{\Phi_0'}{\Phi_0}\,\rho' + \frac{1}{4}\,\frac{\Phi_0''}{\Phi_0}\,\rho\right). \tag{8.50}$$

The equation (8.49) can be solved approximately by a perturbation procedure that essentially is identical to that by which (8.4) was solved. The difference is merely that we replace $\mathcal{R}_3$ in (8.13) by $\mathcal{R}_{ca}$. Further details are then easily worked out.

Other aberrations are due to somewhat more mundane causes such as mechanical defects of electron-optical lenses. We shall not treat those because, as stated in the introductory paragraphs of this chapter, it is increasingly common to simulate electron trajectories on the computer, and both these aberrations and those previously discussed are then obtained explicitly, although numerically. Alternatively, a computer simulation can easily provide numerical studies of aberration coefficients.

REFERENCES

Corbon, H. C. and P. Stehle (1960) *Classical Mechanics*, 2nd edn., John Wiley, New York.

Grivet, P. (1972) *Electron Optics*, 2nd edn., Pergamon, Oxford.

El-Kareh, A. B. and J. C. J. El-Kareh (1970) *Electron Beams, Lenses, and Optics*, Academic Press, Orlando, FL.

Glaser, W. (1952) *Grundlagen der Elektronenoptik*, Springer, Vienna.

Zworykin, V. K., G. A. Morton, E. G. Ramberg, J. Hillier and A. W. Vance (1945) *Electron Optics and the Electron Microscope*, John Wiley, New York.

Born, M. and E. Wolf (1965) *Principles of Optics*, 3rd edn., Pergamon, Oxford.

CHAPTER 9

PHASE SPACE IN ELECTRON OPTICS

9.1 INTRODUCTION

It has become apparent from the analysis in Chapter 3 that an electron in a three-dimensional force field can be described by *six* coordinates in a time-invariant frame. One of these coordinates can be the path parameter $s(t)$ of a reference trajectory. Two others, summarized as a two-dimensional vector $\vec{q}_T$, represent the location of the electron in a (curvilinear) plane characterized by one value of $s(t)$. Two momentum ($\propto$ velocity) coordinates $\vec{p}_T$ describe the orientation of the trajectory at $\vec{r} = [\vec{q}_T(s), s]$ locally along the two $\vec{q}_T$ directions. For this, we also need a sixth momentum coordinate, $p_s(t)$ describing the momentum in the direction that increases path parameter s along that trajectory. In terms of the accelerating potential at the point $\vec{r}$, we have the nonrelativistic conservation theorem,

$$\frac{p_T^2 + p_s^2}{2m} = e\Phi_a(\vec{r}) \,. \tag{9.1}$$

This sixth coordinate is thus derived from the others when we know Φ_a. For the reference trajectory, we have $\vec{p}_T = 0$. Furthermore, the sum $p_T^2 + p_s^2$ is constant in an equipotential plane (if we may ignore the initial momenta so that $\Phi_a \approx \Phi - \Phi_0$); this property will be useful below. Equations (3.58) generalize (9.1) to incorporate magnetostatic $\vec{B}$ fields and relativistic velocities.

The reference trajectory $s(t)$ often can be chosen as a straight line in electrostatics, in which case it can be renamed $z(t)$ so that $\vec{q}_T(s)$ becomes $\vec{\rho}(z)$, and thus a trajectory point is $\vec{r} = [\vec{\rho}(z), z]$. Also, the (nonrelativistic) transverse momentum becomes $\vec{p}_T = m\dot{\vec{\rho}}(z)$. As we observed in Chapter 3, we may consider $z = v_z t = \dot{z}t = p_z t/m$ to show that variable t is linearly related to z in the special case that $p_z(z)$ is a constant. If the reference trajectory is indeed a straight line, then any electron trajectory can be described by the *five* variables $\vec{\rho}(z)$, $\vec{p}(z)$, and $p_z(z)$, which are linked to the fields by means of Eqs. (3.62) with (3.59).

We intend to introduce phase-space considerations (Lichtenberg 1969) in this chapter as an alternative description of charged-particle trajectories. We shall see that phase space has some features that are very helpful in picturing the evolution of beams in electromagnetic fields.

In the most general case, we describe a beam of charged particles (all with the same charge) classically (i.e., nonquantum mechanically) by $3N$ location coordinates $\vec{r}_N = (x_N, y_N, z_N)$ and $3N$ velocity coordinates $\vec{v}_N = (\dot{x}_N, \dot{y}_N, \dot{z}_N)$. The $6N$ coordinates need not be Cartesian; we may replace $\vec{r}_N$, $\vec{v}_N$ by generalized coordinate-momentum variables $\vec{q}_N$, $\vec{p}_N$ as in Chapter 3 (except that we use the letter "q" here for the generalized location coordinates). In actual fact, an N-identical-particle beam is described in classical dynamics by $6N + 1$ variables, $\vec{q}_1 \cdots \vec{q}_N, \vec{p}_1 \cdots \vec{p}_N$, t, which we shall abbreviate by the notation $\vec{q}^{\,3N}$, $\vec{p}^{\,3N}$, t. We then define the particle number density,

$$f_N(\vec{q}^{\,3N}, \vec{p}^{\,3N}, t)\, d^{3N}q\, d^{3N}p \tag{9.2}$$

as the number of particles at time t with location vectors between $\vec{q}_i$ and $\vec{q}_i + d\vec{q}_i$, and momentum vectors between $\vec{p}_i$ and $\vec{p}_i + d\vec{p}_i$ for $1 \leq i \leq N$. It is obvious that

$$\int d^{3N}q \int d^{3N}p\, f_N(\vec{q}^{\,3N}, \vec{p}^{\,3N}, t) = N\,, \tag{9.3}$$

which clarifies that f_N is a particle number density in a $6N$-dimensional space at time t. Division of f_N by N yields a *probability density* of finding the beam in a particular state given by $\vec{q}_1 \cdots \vec{q}_N, \vec{p}_1 \cdots \vec{p}_N$, at time t. It does not matter much here whether we work with particle number density or probability density, and we shall prefer the former.

The beam is characterized by a single point in the above-described $6N$-dimensional phase space at any time t. This point will trace a trajectory in phase space as the beam evolves in time (see Fig. 9-1) where evolution from $t = t_1$ to $t = t_2$ is shown for two beams of N particles each. Figure 9-1 has been simplified to show only one $\vec{q}$ and one $\vec{p}$ coordinate

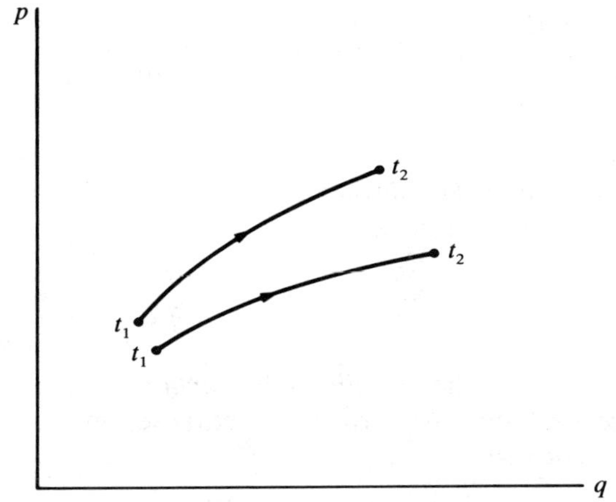

Fig. 9-1. A very schematic sketch of the evolution in time of *two* beams in 6*N*-dimensional phase space.

because we cannot draw more than two or three axes with sufficient clarity, but it suffices to illustrate the situation. At any rate one might consider Fig. 9-1 to be a projection of 6*N* space upon a particular two-dimensional space. The initial conditions for each 6*N*-dimensional trajectory are the 3*N* initial values of both $\vec{q}_i$ and $\vec{p}_i$, i.e., the location coordinates and momenta of the trajectory at $t = t_1$.

The two trajectories in Fig. 9-1 cannot intersect each other because, if they did, the point of intersection would correspond to a solution at that particular time of the equations of motion that would result from *two* distinct sets of initial conditions. The laws of classical mechanics prohibit that; distinct initial conditions lead to distinct trajectories in ordinary space and that implies that at least two of the 6*N* coordinates would be different at all times beyond the initial t_1. Any difference in the 6*N* coordinates implies two distinct points in 6*N* phase space. Thus, if we consider that f_N is a particle density, and that $f_N\dot{\vec{q}}_i$ (or $f_N\dot{\vec{p}}_i$) is a particle flux in 6*N* space, a conservation law is expected to hold (Lichtenberg 1969) in analogy to the conservation-of-charge law taught at introductory electromagnetics levels:

$$\frac{\partial f_N}{\partial z} = -\sum_{i=1}^{3N}\left[\frac{\partial}{\partial \vec{q}_i}\cdot(f_N\dot{\vec{q}}_i) + \frac{\partial}{\partial \vec{p}_i}\cdot(f_N\dot{\vec{p}}_i)\right]. \qquad (9.4)$$

Upon working out the derivatives a bit further, we obtain four sets of terms on the right-hand side:

$$\frac{\partial f_N}{\partial z} = -\sum_{i=1}^{3N} \left(\frac{\partial f_N}{\partial \vec{q}_i} \cdot \dot{\vec{q}}_i + \frac{\partial f_N}{\partial \vec{p}_i} \cdot \dot{\vec{p}}_i \right) - \sum_{i=1}^{3N} f_N \left(\frac{\partial}{\partial \vec{q}_1} \cdot \dot{\vec{q}}_i + \frac{\partial}{\partial \vec{p}_i} \cdot \dot{\vec{p}}_i \right). \tag{9.5}$$

The forces in electron optics can nearly always be derived from a potential, in which case Hamilton's equations (Goldstein 1951, Corbon and Stehle 1960) hold as follows:

$$\dot{\vec{q}}_i = \frac{\partial \mathcal{H}}{\partial \vec{p}_i}, \qquad \dot{\vec{p}}_i = -\frac{\partial \mathcal{H}}{\partial \vec{q}_i}, \tag{9.6}$$

where $\mathcal{H}(\vec{q}^{\,3N}, \vec{p}^{\,3N}, t)$ is the *Hamiltonian function*. These equations have as consequence that the last term in parentheses in (9.5) is zero, and *Liouville's theorem* results:

$$\begin{gathered} \frac{\partial f_N}{\partial t} + \sum_{i=1}^{3N} \left(\frac{\partial f_N}{\partial \vec{q}_i} \cdot \dot{\vec{q}}_i + \frac{\partial f_N}{\partial \vec{p}_i} \cdot \dot{\vec{p}}_i \right) = 0\,, \\ \frac{d}{dt} f_N(\vec{q}^{\,3N}, \vec{p}^{\,3N}, t) = 0\,. \end{gathered} \tag{9.7}$$

The second expression is merely a succinct way of expressing the first; the time evolution of f_N along trajectories is such that its numerical value remains unchanged in the $6N$-dimensional space. It is useful to rewrite (9.7) in yet another way because, in our experience, the interpretation of (9.7) has led to confusion among less experienced readers. Let us suppose that the equations of motion (whatever they are) transform the initial coordinates $\vec{q}_0^{\,3N}, \vec{p}_0^{\,3N}$ at $t = 0$ into the coordinates $\vec{Q}^{3N} = \vec{q}^{\,3N}(t)$, $\vec{P}^{3N} = \vec{p}^{\,3N}(t)$ at time $t > 0$. Considering time interval t to be fixed, we may regard $\vec{Q}^{3N}, \vec{P}^{3N}$ as functions of the initial coordinates $\vec{q}_0^{\,3N}, \vec{p}_0^{\,3N}$. Equation (9.7) can then be re-interpreted to state

$$\begin{aligned} f_N(\vec{Q}^{3N}, \vec{P}^{3N}, t) &= f_N[\vec{Q}^{3N}(\vec{q}_0^{\,3N}, \vec{p}^{\,3N}), \vec{P}^{3N}(\vec{q}_0^{\,3N}, \vec{p}^{\,3N}), t] \\ &\equiv f_N^{(0)}(\vec{q}_0^{\,3N}, \vec{p}_0^{\,3N}, t)\,. \end{aligned} \tag{9.8}$$

The second equality states that we obtain a *new* function of $\vec{q}_0^{\,3N}, \vec{p}_0^{\,3N}$, and t, upon replacing $\vec{Q}^{3N}, \vec{P}^{3N}$ by the functions of $\vec{q}_0^{\,3N}, \vec{p}_0^{\,3N}$ that we obtain upon integrating the equations of motion. The new function $f_N^{(0)}(\vec{q}_0^{\,3N}, \vec{p}_0^{\,3N}, t)$ has the same *numerical* value as $f_N(\vec{Q}^{3N}, \vec{P}^{3N}, t)$. If $\partial f_N/\partial t = 0$, as is often the case in practice, then (9.8) tells us that we can obtain $f_N(\vec{Q}^{3N}, \vec{P}^{3N})$ from $f_N^{(0)}(\vec{q}_0^{\,3N}, \vec{p}_0^{\,3N})$ if this latter function is known (initial beam phase-space density) and if we can work with the inverse equations of motion, i.e., if we can express initial coordinates $\vec{q}_0^{\,3N}, \vec{p}_0^{\,3N}$ as functions of the final coordinates $\vec{Q}^{3N}, \vec{P}^{3N}$.

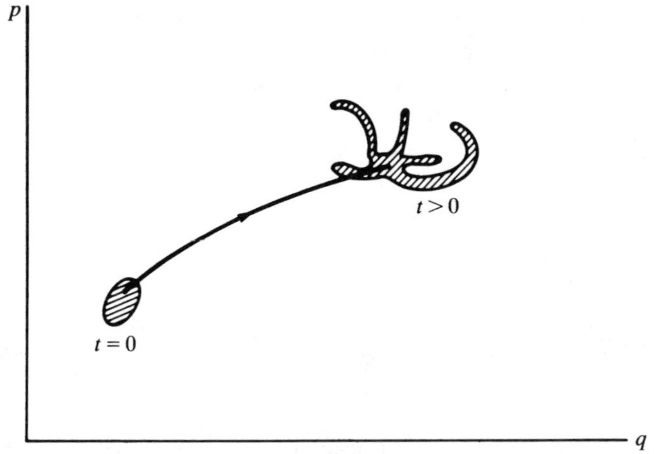

Fig. 9-2. Evolution in time of a beam of particles occupying a region of 6-D phase space. Note distortion of the shape of the region.

Equations (9.7) and (9.8) are manifestations of *Liouville's theorem* (as discussed above), i.e., a central conservation theorem in phase space. The theorem also can be expanded to include nonconservative forces (see Lichtenberg 1969).

Now remember that (9.2) represents a number of particles in a certain $6N$ volume element. This number is obviously conserved along the trajectories from 0 to t which are governed by (9.8). Due to the fact that f_N itself is also conserved, as shown above, it follows that

$$d^{3N}q_0\, d^{3N}p_0 = d^{3N}Q\, d^{3N}P\,, \tag{9.9}$$

i.e., the $6N$-dimensional volume in phase space is also conserved during time evolution of the beam. This is illustrated in Fig. 9-2, which shows that the *shape* of the phase-space volume can be distorted severely even though it is conserved.

9.2 REDUCTION TO SIX-DIMENSIONAL SPACE

It is often more useful to have a form of Liouville's theorem in "ordinary" phase space of six dimensions (three location and three momentum variables). Each of the N particles in a beam of identical particles is represented in this six-dimensional space by one point so that the entire beam is represented at one time by a cluster of N points. Trajectories in this space (which are those of *individual* particles) also cannot cross each other because the unique initial conditions of position and momentum require at least one of the six coordinates to be different for two distinct trajectories in ordinary space.

To obtain a form of Liouville's equation in six-dimensional space we integrate (9.7) over the differential element $dq^{3N-3}\,dp^{3N-3} = d^3q_2 \cdots d^3q_N\,d^3p_2 \cdots d^3p_N$ to obtain

$$\frac{\partial}{\partial t}\left(\int dq^{3N-3} \int dp^{3N-3}\, f_N\right) + \sum_{i=1}^{N}\left(\int dq^{3N-3} \int dp^{3N-3}\, \dot{\vec{q}}_i \frac{\partial f_N}{\partial \vec{q}_i} + \int dq^{3N-3} \int dp^{3N-3}\, \dot{\vec{p}}_i \cdot \frac{\partial f_N}{\partial \vec{p}_i}\right) = 0\,. \tag{9.10}$$

The integral expression in parentheses, of which a $\partial/\partial t$ is taken in (9.10), is only a function of $\vec{q}_1$ and $\vec{p}_1$, and we will name it,

$$f_1(\vec{q}_1, \vec{p}_1) \equiv \int dq^{3N-3} \int dp^{3N-3}\, f_N(\vec{q}^{\,3N}, \vec{p}^{\,3N})\,. \tag{9.11}$$

The other terms in (9.10), considered only for $i = 1$, are easily integrated to yield

$$\dot{\vec{q}}_1 \cdot \frac{\partial f_1}{\partial \vec{q}_1} + \dot{\vec{p}}_1 \cdot \frac{\partial f_1}{\partial \vec{p}_1}\,. \tag{9.12}$$

The remaining summation from $i = 2$ to $i = N$ requires us to look at the terms (for each value of $i > 1$),

$$\int dq^{3N-6} \int dp^{3N-6} \int d^3q_i \int d^3p_i \left(\dot{\vec{q}}_i \cdot \frac{\partial f_N}{\partial \vec{q}_i} + \dot{\vec{p}}_i \cdot \frac{\partial f_N}{\partial \vec{p}_i}\right), \tag{9.13}$$

and the difficulty here is that it is less easy to extract explicit expressions in f_1 from this. The problem is that $\dot{\vec{q}}_i$ is a function of $\vec{p}_i$. If, however, the *forces are derived from a potential*, then the Hamiltonian equations (9.6) hold, and integration of (9.13) by parts yields

$$\begin{aligned}\int dq^{3N-6} \int dp^{3N-6} \sum_{\alpha=1}^{3} \int d^3q_i \int d^3p_i \left(\frac{\partial \mathcal{H}}{\partial p_{i\alpha}} \frac{\partial f_N}{\partial q_{i\alpha}} - \frac{\partial \mathcal{H}}{\partial q_{i\alpha}} \frac{\partial f_N}{\partial p_{i\alpha}}\right) \\ = \int dq^{3N-6} \int dp^{3N-6} \sum_{\alpha=1}^{3} \left[\int d^3q_i \int_c d^2p_i\, f_N(p_{i\alpha}) \frac{\partial \mathcal{H}}{\partial p_{i\alpha}} \right. \\ \left. - \int_c d^2q_i \int d^3p_i\, f_N(q_{i\alpha}) \frac{\partial \mathcal{H}}{\partial q_{i\alpha}}\right],\end{aligned} \tag{9.14}$$

where the subscript "c" on the two-dimensional integral signs indicates a closed-surface integration, and where the dependence upon the relevant surface coordinate of f_N is explicitly shown. Because $q_{i\alpha}$ and $p_{i\alpha}$ are on an infinitely distant surface where they tend towards infinity, and because $f_N \to 0$ when any of its position or momentum variables tend towards infinity, it follows that all the terms in (9.14) are zero. The net result for reduction of (9.10) is

$$\frac{\partial f_1}{\partial t} + \dot{\vec{q}} \cdot \frac{\partial f_1}{\partial \vec{q}} + \dot{\vec{p}} \cdot \frac{\partial f_1}{\partial \vec{p}} = 0 , \tag{9.15}$$

where $\vec{q}$ and $\vec{p}$ are two position and momentum vectors. Equation (9.15) is the formulation of Liouville's theorem in the six-dimensional (6-D) space spanned by the three position and the three momentum coordinate of a single particle, and $f_1(\vec{q}, \vec{p})$ is the single-particle number density in this space which yields N (the number of particles) after integration of all six variables over the entire 6-D space.

To state it again, a sufficient condition for the reduction of (9.4) to (9.15) is that the forces can be derived from a potential. This may seem to exclude a beam of charged particles because the Coulomb forces introduce pairwise forces between two charged particles. However, as we shall discuss more fully in Chapter 10, a beam of charged particles may be considered, to very good approximation, to consist of "independent" particles, each of which exists in a central Coulomb field produced by the beam as a whole. The central Coulomb field can be derived from a potential which is a function only of the local position variables, and that is a sufficient condition for the validity of (9.15). Lichtenberg (1969) reviews more general conditions under which other (noncentral) interparticle forces can lead to a valid (9.15).

The 6-D formulation of Liouville's theorem, Eq. (9.15), can also be written as

$$\frac{df_1}{dt} = 0 , \qquad f_1[\vec{Q}(\vec{q}, \vec{p})\, \vec{P}(\vec{q}, \vec{p})] = f_1^{(0)}(\vec{q}, \vec{p}) , \tag{9.16}$$

in analogy to (9.8). To state it again, $f_1(\vec{q}, \vec{p})$ is a single-particle distribution density that yields the total number of particles in a beam when integrated over its six variables. Equations (9.16) express the conservation of f_1 as the trajectory coordinates $\vec{q}$, $\vec{p}$ evolve in time into $\vec{Q}$, $\vec{P}$. It is possible that f_1 depends explicitly upon time but we will introduce that in our notation only when needed. We also have conservation of 6-D phase-space volume,

$$\frac{d}{dt}(d^3q\, d^3p) = 0\,, \tag{9.17}$$

because $f_1(\vec{q}, \vec{p})\, d^3q\, d^3p$ represents the number of particles with coordinate values between $\vec{q}, \vec{p}$ and $\vec{q} + d\vec{q}, \vec{p} + d\vec{p}$ and because f_1 itself is conserved.

9.3 REDUCTION TO FOUR-DIMENSIONAL SPACE

Particle beams can be characterized with respect to a *centroid* (i.e., a convenient reference trajectory such as, possibly, a central one) for many, if not most, cases of interest. For example, charged particles with one and the same accelerating potential $\Phi_a(\vec{r}_0)$ on an initial surface will often travel in time t to another well-defined surface; usually an equipotential surface. These surfaces are planar outside the lensing region of axisymmetric (cylindrically symmetric) electron lenses. More precisely, the particles will travel to an equipotential surface after an equal expenditure of energy $[\delta(\frac{1}{2}mv^2) = \delta(e\Phi)]$ whenever the electromagnetic forces can be derived from a potential, and that is almost always the case in electron optics. Every incremental change $\delta(\frac{1}{2}mv^2)$ in kinetic energy implies an incremental change δv that is identical for each particle in the beam. Then, because $ds = v(t)\, dt + t\, dv$ in going from t, s to $t + dt, s + ds$, we observe that fixed dt and fixed $v(t)$ yields fixed $s(t + dt)$ and fixed $v(t + dt)$ along each trajectory, i.e., the new equipotential plane after dt lies at fixed ds for all trajectories. Hence, if the equipotential surfaces are known from calculations of potential and fields (Chapter 4), then a *single* reference trajectory path coordinate $s(t)$ will suffice to define the relevant equipotential surface, and $v(s)$ follows from (3.8). As a consequence, we need only to know the transverse coordinates q_1, p_1, q_2, p_2 for each particle.

However, if the electrons have a distribution of *thermal* energies in their initial state at $s = 0$, then the above argument is only approximate because the kinetic energies will differ by $\frac{1}{2}mv_{th}^2$ (if v_{th} is the thermal velocity of an individual electron at its source). The accelerating potential is given by

$$e\Phi_a = e(\Phi - \Phi_0) + \tfrac{1}{2}mv_{th}^2\,, \tag{9.18}$$

and in most practical applications (in cathode-ray tubes), the thermal kinetic energy term is extremely small compared to the potential difference except in a very small region near the source. To state the argument somewhat more precisely, let us remember that

$$\tfrac{1}{2}mv^2(t) - e\Phi(t) = \tfrac{1}{2}mv^2(0) - e\Phi(0)\,, \tag{9.19}$$

along each trajectory, and that $p_3(t) = mv(t)$. We define $q_1(q_3)$ and $q_2(q_3)$ by placing equipotential surfaces at fixed locations $s = q_3(t)$ along the centroid and by defining q_1 and q_2 as (curvilinear) transverse coordinates that measure the distance along two mutually orthogonal directions in the equipotential plane to the reference trajectory. The transverse momenta in these directions are p_1 and p_2. Let us write down the integral expression for each $q_i(t)$ with $i = 1, 2$ using the equations of motion that follow from (3.2) in the absence of a magnetic field:

$$\begin{aligned}
q_i(t) &= q_i(0) + \int_0^t dt_1\, \dot{q}_i(t) = q_i(0) + \int_0^{Q_3} dq_3\, \frac{\dot{q}_i(q_3)}{dq_3/dt} \\
&= q_i(0) + \int_0^{Q_3} dq_3\, \frac{m\dot{q}_i(q_3)}{m\, dq_3/dt} = q_i(0) + \int_0^{Q_3} dq_3\, \frac{p_i(q_3)}{p_3(q_3)}\,, \\
p_i(t) &= p_i(0) + \int_0^t dt_1\, \dot{p}_i(t) = p_i(0) - \int_0^t dt_1\, E_i(t_1) \\
&= p_i(0) - e\int_0^{Q_3} dq_3\, \frac{E_i(q_3)}{dq_3/dt} = p_i(0) - em\int_0^{Q_3} dq_3\, \frac{\partial\Phi(q_3)/\partial q_i}{p_3(q_3)}\,, \\
|\vec{p}(q_3)|^2 &= 2em[\Phi(q_3) - \Phi(0)] + |\vec{p}(0)|^2\,.
\end{aligned} \tag{9.20}$$

We will abbreviate $\vec{q}(t)$ by $\vec{Q}$ and $\vec{p}(t)$ by $\vec{P}$ in these equations, i.e., $q_i(q_3) = Q_i$ and $p_i(q_3) = P_i$. The integrations are along trajectory paths so that, for example, $E_i(q_3)$ is an abbreviated notation for the value of the electrostatic field at the trajectory point in question with $q_3(t)$ as its reference coordinate. We have assumed a very small spread in initial thermal energy, i.e., the transverse $p_i(0)$ are very small compared to $\sqrt{2em[\Phi(Q_3) - \Phi(0)]}$, and also that $p_3(t) \gg p_3(0)$ for the reference trajectory. It then follows from the fifth line of (9.20) that

$$p_3^2(q_3) \approx 2em[\Phi(q_3) - \Phi(0)]\,, \tag{9.21}$$

for all trajectory locations where $\Phi(q_3)$ has become sufficiently large so that $2em[\Phi - \Phi_0] \gg p_3^2(0)$. This is usually true in the lensing regions of electron optics lenses (where $\partial\Phi(q_3)/\partial q_i \neq 0$). The transverse-momentum equation in (9.20) states that P_i differs from $p_i \equiv p_i(0)$ by a momentum-change integral which is also given in (6.15) in terms of more familiar lens quantities. The point to be made here is that the momentum *change* is of the same order of magnitude as the momentum itself before change. Therefore, the thermal spread in P_i will be of the same order of magnitude as that in p_i, although the *central* value of p_i may change according to (6.15).

To summarize, p_1, p_2 and p_3, at time $t = 0$, may exhibit a thermal spread (often of several tenths of an electron volt), and P_1, P_2 will show a shifted thermal spread of essentially similar width around some new value of the same order of magnitude. P_3, however, will be given by (9.21) approximately at $q_3 = Q_3$, and it will be orders of magnitude larger than p_3 in actual accelerating lenses. The ratio $\delta P_3/P_3$, which is of the order of $\delta p_3/P_3$ will thus tend to be extremely small compared to unity; the thermal spread in the longitudinal q_3 direction manifests itself as a small *chromatic* aberration which can be ignored in many applications. As a result of the above considerations, we usually can approximate the 6-D single-particle density as follows,

$$f_1(\vec{q}_T, q_3, \vec{p}_T, p_3) = \bar{f}_1(\vec{q}_T, \vec{p}_T; q_3)\delta(p_3 - \bar{p}_3)\,, \tag{9.22}$$

where $\bar{p}_3$ is the mean value of the thermally distributed longitudinal momentum coordinate $p_3(q_3)$ at time t. The subscript "T" signifies the two transverse coordinates. Hence, at $q_3(t)$ we have the beam current,

$$\begin{aligned} I \equiv I(q_3) &= \frac{1}{m}\int d^2q_T \int d^2p_T \int_{-\infty}^{\infty} dp_3 p_3\, f_1(\vec{q}_T, q_3, \vec{p}, p_3) \\ &= \frac{1}{m}\int d^2q_T \int d^2p_T\, \bar{f}_1(\vec{q}_T, \vec{p}_T; q_3)\,. \end{aligned} \tag{9.23}$$

The distribution of $p_3(0)$ at time $t = 0$ over the various beam trajectories is usually not sharply peaked, and specifically,

$$I = I(0) = \frac{1}{m}\int d^2q_T \int d^2p_T \int_{-\infty}^{\infty} dp_3 p_3\, f_1(\vec{q}_T, 0, \vec{p}_T, p_3)\,. \tag{9.24}$$

We wish to make this equation for the source beam current resemble (9.23) more closely. To do so, we note that if $f_1(\vec{q}_T, 0, \vec{p}_T, p_3)$ is a known function (either analytically or numerically) then we can define a "mean" value of p_3 as

$$\bar{p}_{30} \equiv \frac{\int d^2q_T \int d^2p_T \int_{-\infty}^{\infty} dp_3 p_3\, f_1(\vec{q}_T, 0, \vec{p}_T, p_3)}{\int d^2q_T \int d^2p_T \int_{-\infty}^{\infty} dp_3\, f_1(\vec{q}_T, 0, \vec{p}_T, p_3)}\,, \tag{9.25}$$

to obtain at the source,

$$I = I(0) = \frac{1}{m}\, \bar{p}_{30} \int d^2q_T \int d^2p_T \int_{-\infty}^{\infty} dp_3\, f_1(\vec{q}_T, 0, \vec{p}_T, p_3)\,. \tag{9.26}$$

To finish this, define

$$\bar{f}_1(\vec{q}_T, \vec{p}_T; 0) \equiv \int_{-\infty}^{\infty} dp_3\, f_1(\vec{q}_T, 0, \vec{p}_T, p_3)\,, \tag{9.27}$$

in order to obtain for the source beam current,

$$I = I(0) = \frac{1}{m}\, \bar{p}_{30} \int d^2q_T \int d^2p_T\, \bar{f}_1(\vec{q}_T, \vec{p}_T; 0)\,. \tag{9.28}$$

The expressions for $I(q_3)$ in (9.23) and $I(0)$ in (9.28) yield the same numerical value for the beam current because current is conserved. It serves our purpose well to define a four-dimensional (4-D) *current distribution density*,

$$F(\vec{q}_T, \vec{p}_T; q_3) \equiv \frac{1}{m}\, \bar{p}_3 \bar{f}_1(\vec{q}_T, \vec{p}_T; q_3)\,, \tag{9.29}$$

because the conservation of beam current then states explicitly

$$\frac{d}{dq_3}\,[F(\vec{q}_T, \vec{p}_T; q_3)] = 0\,. \tag{9.30}$$

In most practical situations, $F(\vec{q}_T, \vec{p}_T)$ will depend implicitly through $\vec{q}_T(q_3)$, $\vec{p}_T(q_3)$ upon the path variable q_3, rather than explicitly, and then (9.17) yields

$$\frac{d}{dq_3}\,(d^2q_T\, d^2p_T) = 0\,, \tag{9.31}$$

as a statement on the conservation of the 4-D phase-space volume.

9.4 TRANSFORMATIONS IN FOUR-DIMENSIONAL SPACE

Our discussions of phase space have now led to a description of a charged-particle beam in an accelerating lens by means of a four-dimensional particle-flux or current distribution density $F(\vec{q}_T, \vec{p}_T; q_3)$ that is conserved as a longitudinal path parameter $q_3(t)$ progresses in time. Furthermore, the 4-D volume element $d^2q_T\, d^2p_T$ in phase space is conserved along the particle trajectories. Let us now specialize for the sake of simplicity to an axisymmetric beam so that $q_3 \equiv z$. The volume element becomes $d^2q_T\, d^2p_T = 4\pi^2 q_T\, dq_T p_T\, dp_T = \pi^2 d(q_T^2)\, d(p_T^2)$. We may identify $d(q_T^2)$ with ρ^2 and $d(p_T^2)$ with the transverse momentum p_T^2 when the volume element is applied to a small paraxial bundle of particle trajectories around the z axis, so that we conclude that $\rho^2(z)p_T^2(z)$ is conserved along z. However, $p_T^2(z) = p^2(z)\sin^2\theta = 2em\Phi_a \sin^2\theta$, in the understanding that Φ_a is the accelerating potential (which can be approximated by the spatial potential Φ if indeed $\Phi \gg \Phi_0$). The volume conservation (9.31) translates into

$$\frac{d}{dz}(\sqrt{\Phi}\rho \sin\theta) = 0\,, \tag{9.32}$$

which is the Abbe sine law. Compare the derivation of (9.32) to that of Eq. (6.8) which is paraxial because tan θ rather than sin θ appears (tan θ and sin θ are approximately equal for small θ); the present point of view seems to be somewhat broader. In the case of planar symmetry we have $dy = 0$ and $dp_y = 0$ so that xp_x is conserved, yielding

$$\frac{d}{dz}(\sqrt{\Phi}x \sin\theta) = 0\,. \tag{9.33}$$

We will now derive some simple matrix transformations for elementary situations from the more accurate expressions,

$$\frac{d}{dz}[\rho(z)p_T(z)] = 0\,. \tag{9.34}$$

The canonical form of all of these transformations is

$$\begin{bmatrix} \rho \\ p_T \end{bmatrix} = \mathcal{T}\begin{bmatrix} \rho_0 \\ p_{T0} \end{bmatrix} = \begin{bmatrix} t_{11} & t_{12} \\ t_{21} & t_{22} \end{bmatrix}\begin{bmatrix} \rho_0 \\ p_{T0} \end{bmatrix}, \tag{9.35}$$

where the determinant of the transformation matrix $\mathcal{T}$ must be unity in order that (9.34) is satisfied. Complicated lenses often can be described as a series of *drift* and *momentum-change* transformations. Let us consider these separately in the following sections.

9.4.1 Drift Transformations

The simplest *drift* transformation describes the linear motion of an electron in a purely accelerating field without transverse forces ($\vec{E}_T = 0$); there is no change in transverse momentum. We define the matrix

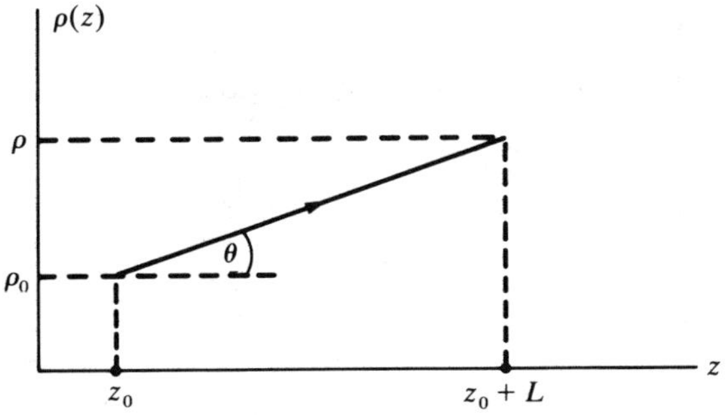

Fig. 9-3. A field-free trajectory that gives rise to an elementary drift transformation in phase space.

transformation from ρ_0, p_{0T} to ρ, p_T, with reference to Fig. 9-3, by $\mathscr{D}$ with

$$\mathscr{D} = \begin{bmatrix} 1 & \dfrac{L}{p_z} \\ 0 & 1 \end{bmatrix}. \tag{9.36}$$

9.4.2 Momentum-Change Transformations

Recalling (6.15) from what we have written about Gaussian optics, we observe that that equation can be rewritten, upon defining the transverse momentum by the approximation, $p_T \approx \sqrt{2em\Phi}\rho'$, as

$$p_T - p_{0T} = -\rho_L\left[\sqrt{2em}\,\frac{1}{4}\int_{L-\epsilon}^{L+\epsilon} d\zeta\, \frac{\Phi_0''(\zeta)}{\sqrt{[\Phi_0(\zeta)]}}\right] = -S\rho_L\,. \tag{9.37}$$

We may consider this as an approximation either to a *thin* lens (in which case ϵ is very small) or to a *weak* lens (in which case the variations in Φ are small). In both cases $\rho(\zeta)$ in the integrand is approximated reasonably well by the constant ray height ρ_L. We identify $S = \sqrt{2em}\, s$, where s is the strength factor discussed in connection with (6.15).

The momentum-transfer described by (9.37) is depicted in Fig. 9-4 (in which both principal planes have been merged into the single one shown at $z = L$). It then becomes clear that ρ_0, p_{0T} transform into ρ, p_T by means of a matrix $\mathscr{M}$,

$$\mathscr{M} = \begin{bmatrix} 1 & 0 \\ -S & 1 \end{bmatrix}. \tag{9.38}$$

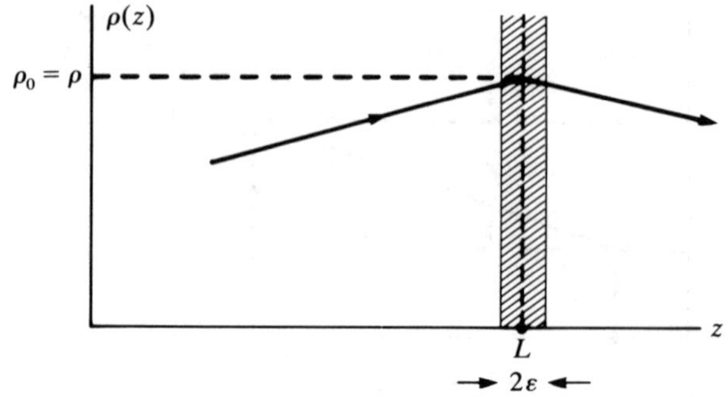

Fig. 9-4. Transition of an electron through a thin lens, corresponding to a momentum-change transformation in phase space.

Clearly, $\mathcal{M}$ as well as $\mathcal{D}$ has unit determinant so that any product of $\mathcal{M}$ and $\mathcal{D}$ matrices (in any order) also has unit determinant, thus ensuring that the phase-space conservation law (9.34) is satisfied.

Let us consider a simple example to illustrate the above, i.e., the focussing of principal electrons as sketched in Fig. 9-5. Initially there is a drift (with $p_{0T} = 0$) from $z = 0$ to $z = L$, then there is a momentum change at $z = L$, and finally there is a drift from $z = L$ to $z = L + f$. In terms of the sketch,

$$\begin{bmatrix} \rho \\ p_T \end{bmatrix} = \begin{bmatrix} 1 & -\dfrac{f}{p_z} \\ 0 & 1 \end{bmatrix} \begin{bmatrix} 1 & 0 \\ -S & 1 \end{bmatrix} \begin{bmatrix} 1 & \dfrac{L}{p_{z0}} \\ 0 & 1 \end{bmatrix} \begin{bmatrix} \rho_0 \\ 0 \end{bmatrix}. \tag{9.39}$$

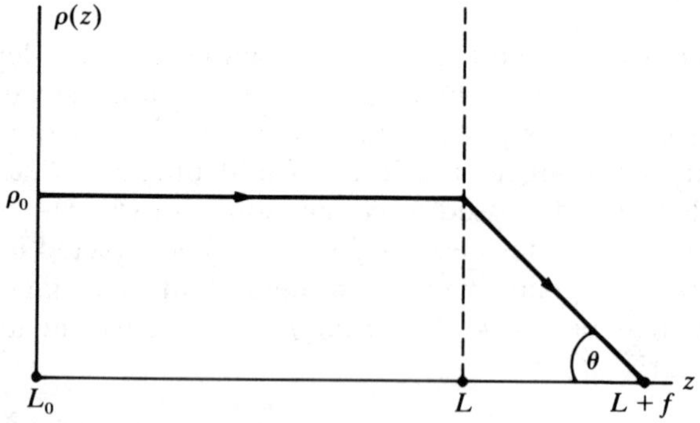

Fig. 9-5. Focussing of principal electrons, corresponding to simple compound drift and momentum-change transformations in phase space.

The right-most 2×2 matrix has no effect upon ρ_0 and $p_{0T} = 0$, hence it may be ignored. The matrix equation then becomes,

$$\begin{bmatrix} \rho \\ p_T \end{bmatrix} = \begin{bmatrix} 1 + \dfrac{Sf}{p_z} & -\dfrac{f}{p_z} \\ -S & 1 \end{bmatrix} \begin{bmatrix} \rho_0 \\ 0 \end{bmatrix}, \tag{9.40}$$

and from this we learn (again!) that $p_T = -S\rho_0$ as is given by (9.37) explicitly. Equation (9.40) also yields that $\rho = [1 + (Sf/p_z)]\rho_0$, which in combination with $S = -p_T/\rho_0$, yields $\rho = \rho_0 - fp_T/p_z = \rho_0 - f \tan\theta = 0$, just as one might expect in the focal plane! The reader should have noticed that we read matrix multiplication from *right* to *left* as the electron moves from left to right.

As another example, consider the general thin- or weak-lens approximation, as depicted in Fig. 9-6. It can be described by two drifts sandwiching one momentum change:

$$\begin{bmatrix} \rho \\ p_T \end{bmatrix} = \begin{bmatrix} 1 & \dfrac{L}{p_z} \\ 0 & 1 \end{bmatrix} \begin{bmatrix} 1 & 0 \\ -S & 1 \end{bmatrix} \begin{bmatrix} 1 & \dfrac{L_0}{p_{z0}} \\ 0 & 1 \end{bmatrix} \begin{bmatrix} \rho_0 \\ 0 \end{bmatrix}. \tag{9.41}$$

After multiplying out the three 2×2 matrices this becomes

$$\begin{bmatrix} \rho \\ p_T \end{bmatrix} = \begin{bmatrix} 1 - \dfrac{LS}{p_z} & \dfrac{L_0}{p_{z0}} + \dfrac{L}{p_z} - \dfrac{L_0 LS}{p_{z0} p_z} \\ -S & 1 - \dfrac{L_0 S}{p_{zo}} \end{bmatrix} \begin{bmatrix} \rho_0 \\ p_{0T} \end{bmatrix}. \tag{9.42}$$

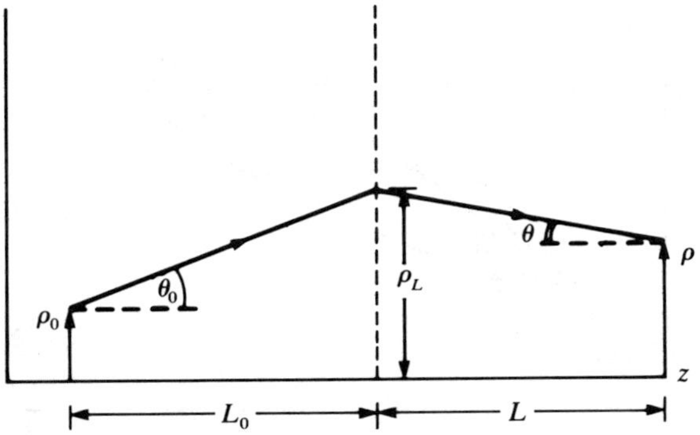

Fig. 9-6. Imaging in the thin-lens approximation, corresponding to another simple compound phase-space transformation.

We would have focussing if $p_{0T}=0$. Indeed, we then would find $\rho = [1-(LS/p_z)]\rho_0$ and $p_T = -S\rho_0$. We found this second relationship already from (9.40). Replacing S in the first relationship, we obtain $\rho = \rho_0 + (Lp_T/p_z) = \rho_0 - L\tan\theta$, which is precisely what can be read from the geometry in Fig. 9-6. If, on the other hand, we regard an *imaging* situation ($\rho_0 = \rho = 0$), then we obtain from (9.42)

$$p_T = \left(1 - \frac{L_0 S}{p_z}\right) p_{0T}\,, \qquad \frac{L_0}{p_{0z}} + \frac{L}{p_z} = \frac{L_0 L S}{p_{0z} p_z}\,. \tag{9.43}$$

Because $p_{0T}/p_{0z} = \tan\theta$, it follows that the first formula in (9.43) translates into $p_T - p_0 = -S\rho_L$, which is (9.37). The remaining equation essentially yields the Lagrange-Helmholtz relationship (6.12) or (6.16),

$$\frac{p_{0z}}{L_0} + \frac{p_z}{L} = S\,. \tag{9.44}$$

In fact, almost all of paraxial optics is easily reconstructed from combinations of matrices $\mathscr{D}$ and $\mathscr{M}$ as products. All paraxial lensing is constrained by the conservation laws so easily expressed in phase space.

9.4.3 Beam Emittance, Acceptance

Consider a beam described by the current distribution density in 4-D phase space, $F(\vec{\rho}, \vec{p}_T; z)$. We have sketched a region in Fig. 9-7 with a sharp contour in an (r, p) space as a 2-D representation of a four-dimensional volume at $z=0$. We assume $F(\vec{\rho}_0, \vec{p}_{0T}; 0) = 0$ outside the

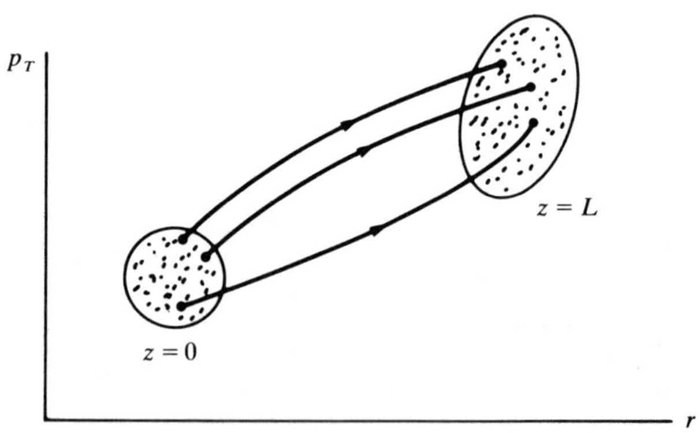

Fig. 9-7. Beam evolution in 4-D phase space to illustrate the concepts of acceptance and emittance.

contour. The area (volume) inside the contour is known as the *emittance*, and it can be written as

$$\mathbb{A}_e = \int\int d^2\rho_0 \, d^2 p_{0T} = p_{0z}^2 \int d^2\rho_0 \, \Omega_0(x_0, y_0) , \tag{9.45}$$

if $\Omega_0(x, y)$ is the solid angle between the extreme values of p_{0x} and p_{0y} at fixed x_0, y_0. Similarly, at $z = L$ we have the *acceptance*,

$$\mathbb{A}_a = \int\int d^2\rho \, d^2 p = p_z^2 \int d^2\rho \, \Omega(x, y) . \tag{9.46}$$

Further discussion of these concepts are given by Dahl (1973). The concepts are probably most useful when there is a relatively sharp transition at the contour from very low densities to a more or less uniform peak density, in which case it spares us the necessity of dealing with $F(\vec{\rho}, \vec{p}_T; z)$. Modern phase-space techniques, however, enable us to work effectively with the current-density function $F(\vec{\rho}, \vec{p}_T; z)$.

REFERENCES

Born, M. and E. Wolf (1965) *Principles of Optics*, 3rd edn., Pergamon, Oxford.

Corbon, H. C. and P. Stehle (1960) *Classical Mechanics*, 2nd edn., John Wiley, New York.

Dahl, P. (1973) *Introduction to Electron and Ion Optics*, Academic Press, New York.

Goldstein, H. (1951) *Classical Mechanics*, Addison-Wesley, Reading, MA.

Kerr, D. E. (Ed.) (1951) *Propagation of Short Radio Waves*, McGraw-Hill, New York.

Lichtenberg, A. J. (1969) *Phase-Space Dynamic of Particles*, John Wiley, New York.

CHAPTER 10

TREATMENT OF SPACE CHARGE

A major difficulty in calculating the propagation of electron beams (through lenses, etc.) lies in the treatment of the mutually repulsive Coulomb forces of the electrons. The same is true for other charged particles, but we will restrict the discussion to electrons which are more easily deflected due to their relatively small mass.

Consider one electron in a beam of electrons under influence of an electrostatic field $\vec{E}$. Its nonrelativistic equation of motion is

$$\ddot{\vec{r}}_i = -\frac{e}{m}\left[\vec{E} + \sum_{j\neq i} \frac{q_j(\vec{r}_i - \vec{r}_j)}{4\pi\epsilon_0|\vec{r}_i - \vec{r}_j|^3}\right]. \tag{10.1}$$

Referring to Eq. (3.19), we can rewrite the transverse part of this vector equation in terms of an independent z variable,

$$\vec{\rho}_i'' = -\frac{1+\vec{\rho}_i'^2}{2\Phi_a}\left[\vec{E}_\rho + \vec{\rho}_i' E_z + \sum_{j\neq i} \frac{q_j(\vec{\rho}_i - \vec{\rho}_j)}{4\pi\epsilon_0|\vec{r}_i - \vec{r}_j|^3}\right]. \tag{10.2}$$

The space-charge problem ("space charge" is the usual name given to the Coulomb-force aspects of charged-particle beams) can be understood as follows. Most numerical schemes for solving Eq. (10.2) derive from difference equations such as

$$\begin{aligned} \vec{\rho}_i(z+dz) &= \vec{\rho}_i(z) + \vec{\rho}_i'(z)dz + \tfrac{1}{2}\rho_i''(z)(dz)^2 + \cdots \\ \vec{\rho}_i'(z+dz) &= \vec{\rho}_i'(z) + \vec{\rho}_i''(z)dz + \cdots . \end{aligned} \tag{10.3}$$

Integration based upon Eq. (10.3) is known as *Euler's method*, and although it is overly simplistic, it serves to illustrate the principles involved. Numerical integration schemes for total differential equations are usually more sophisticated; e.g., Runge-Kutta methods are often preferable (James et al. 1985). The emphasis here is on the usefulness of "marching" methods; when $\vec{\rho}_i(z)$ is known, we should be able to estimate $\vec{\rho}_i(z+dz)$ and thus gradually "march" the solution from $z=0$ to $z=L$. If the summation term were missing in Eq. (10.2) then use of that equation at z would immediately yield $\vec{\rho}_i(z+dz)$ and $\vec{\rho}_i'(z+dz)$ by means of Eq. (10.3) at least up to second order in dz. However, the summation term involves values of $\vec{\rho}$ at points beyond z; points which have not yet been reached! One way of overcoming this complication is to take recourse to an *iteration* procedure in solving Eq. (10.2). An overview of the procedure is as follows:

Step 1: Set the summation term in Eq. (10.2) equal to zero, and solve (10.2) by means of Eq. (10.3) or by more sophisticated methods that "march" the initial values from $z=0$ to $z=dz$, then to $z=2dz$, etc., until $z=L$ is reached.

Step 2: Use the thus obtained zero-order trajectories $\vec{\rho}_i^{(0)}(z)$ to form the space-charge term,

$$\vec{E}_{sp}^{(1)} = \sum_{j\neq i} \frac{q_j(\vec{\rho}_i^{(0)} - \vec{\rho}_j^{(0)})}{4\pi\epsilon_0|\vec{r}_i^{(0)} - \vec{r}_j^{(0)}|^3} . \tag{10.4}$$

Step 3: Recalculate the trajectories from Eq. (10.3), with Eq. (10.4) replacing the summation term of Eq. (10.2), to obtain new coordinates $\vec{\rho}_i^{(1)}(z)$ and $d\vec{\rho}_i^{(1)}/dz$.

Iteration procedure: Repeat Steps 2 and 3 until the difference between the $(n+1)$st- and nth-order trajectories is negligible.

The numerical complexity of this approach is quite apparent. Not only must we compute many trajectories for a beam (a Monte-Carlo approach, for example, requires a very appreciable number for accuracy) but we must recompute them many times! Therefore, approximations to Eq. (10.4), the space-charge field, that can circumvent this extra complication are of vital importance for practical calculations of beam propagation.

We may consider Eq. (10.4) to be approximated well at $\vec{r}_i = \vec{r}$ by

$$\begin{aligned} \vec{E}(\vec{r}) &= -\vec{\nabla}\Phi(\vec{r}) , \\ \Phi(\vec{r}) &= \frac{1}{4\pi\epsilon_0} \int\!\!\int\!\!\int d^3r_1 \, \frac{\omega(\vec{r}_1)}{|\vec{r} - \vec{r}_1|} , \end{aligned} \tag{10.5}$$

where $\omega(\vec{r}_1)$ is the volume-charge density at $\vec{r}_1$. In the static-beam situation, $\omega(\vec{r}_1)$ is not a function of time. In practical time-dependent situations all changes with time will be sufficiently slow for a quasi-static approximation; Eq. (10.5) is valid, but $\Phi(\vec{r}, t)$ may well change in time, and induced magnetic effects are negligible. The continuity of $\omega(\vec{r}_1)$ ensures that the apparent singularity at $\vec{r}_1 = \vec{r}$ is not felt. That was alluded to above, where we mentioned a brute-force Monte-Carlo "shot-gun" approach to trajectory simulation. Such an approach has great difficulties in dealing with the relative unimportance of the small volume element d^3r_1 in the immediate vicinity of $\vec{r}_1 = \vec{r}$. In the Monte-Carlo approach this small volume element is sampled infrequently which then results in large errors because the $(1/r)$-type singularity is not balanced out. For this reason alone, a Monte-Carlo approach to trajectory simulation is undesirable. Of course Eq. (10.5) is the integral form of *Poisson's equation*,

$$\Delta\Phi(\vec{r}) = -\frac{\omega(\vec{r})}{\epsilon} . \tag{10.6}$$

We shall now deal with various important space-charge effects directly. We mainly highlight the essential features and refer to Ivey (1954) for further details.

10.1 THE LAMINAR UNIFORM AXISYMMETRIC BEAM

A classical canonical problem involving space charge is the analytical calculation of a beam of electrons in which the initial charge density $\omega(\vec{\rho}, 0)$ is uniform in $\vec{\rho}$ at $z = 0$, i.e., $\omega(\vec{\rho}, 0) = \omega(0)$, and in which all velocities at $z = 0$ are parallel and identical. This gives rise to a *laminar* beam (in the absence of lensing effects) because trajectories do not cross each other anywhere. The space charge field in an axisymmetric beam is governed by the Maxwell div-$\vec{E}$ equation in cylindrical coordinates:

$$\frac{1}{\rho}\frac{\partial}{\partial\rho}(\rho E_\rho) + \frac{\partial E_z}{\partial z} = \frac{1}{\epsilon_0}\,\omega(\rho, z) . \tag{10.7}$$

We will make several important assumptions in solving this equation. These assumptions will be justified a posteriori in a self-consistent fashion, i.e., we will show that the solutions so obtained do indeed satisfy the assumptions.

(i) The field due to the accelerating potential is in the $\hat{z}$ direction and it greatly exceeds the space-charge field.

(ii) The longitudinal space-charge field is negligible.

(iii) The current density along the ray trajectory $\rho(z)$, namely $J(s)$, is uniform in a (curved) plane perpendicular to the trajectory rays characterized by parameter s with line element $ds = \sqrt{d\rho^2(z) + dz^2}$.

Let us consider curvilinear orthogonal coordinates $\vec{\tau}$, s for the trajectory at s such that $\vec{\tau}(s)$ is the transverse-coordinate vector to the main coordinate s which is essentially a path parameter for each trajectory. A useful choice of coordinates will have equipotential planes defined by $s = \text{constant}$ planes. In the case of axisymmetric systems there will be no dependence of meridional trajectories on the azimuthal angle and the use of a path parameter on a slowly curving trajectory allows us more accuracy than would be obtained by using $\rho(z)$, z coordinates. The integral form of Eq. (10.7) is then (see Fig. 10-1)

$$2\pi \int_0^{\tau} d\tau_1 \tau_1 \,[E_s(\tau_1, s + ds) - E_s(\tau_1, s)] + 2\pi ds \tau E_\tau(\tau, s)$$

$$= \frac{2\pi}{\epsilon_0} ds \int_0^{\tau} d\tau_1 \tau_1 \,\omega(\tau_1, s)\,. \tag{10.8}$$

Assumption (ii) allows us to neglect the first term in Eq. (10.8). By definition, $\omega(\tau, s) = -J(s)/v(s)$ in terms of current density and velocity, and therefore is not a function of τ to the extent that $v(s)$ is uninfluenced in magnitude by τ when the longitudinal space-charge field is ignored (assumption (ii)). Under these circumstances, we obtain

$$E_\tau(\tau, s) = \frac{1}{\epsilon_0 \tau} \int_0^{\tau} d\tau_1 \tau_1 \,\omega(s) = -\frac{J(s)}{2\epsilon_0 v(s)}\,\tau\,. \tag{10.9}$$

The transverse-force equation $m\ddot{\vec{\tau}} = -e\vec{E}_\tau$ becomes

$$\frac{d^2\tau}{ds^2} = -\frac{1}{2\Phi_a} E_\tau(\tau, s) = \frac{1}{4\epsilon_0} \sqrt{\frac{m}{2e}}\, \frac{J(s)}{\Phi_a^{3/2}}\, \tau(s)\,, \tag{10.10}$$

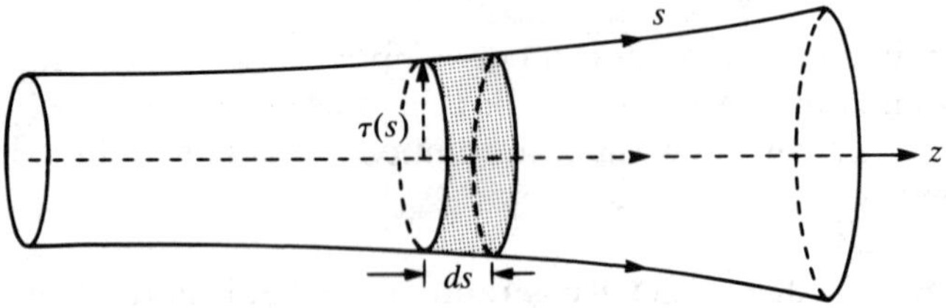

Fig. 10-1. A portion of a laminar uniform axisymmetric beam.

where we have used Eq. (3.21) to replace $v(s)$ by $\Phi_a^{1/2}$. We define $I_\tau \equiv J(s)\pi\tau^2(s)$ as the current enclosed by the trajectory with axisymmetric coordinates $[\tau(s), s]$. It is not a function of s due to the laminar nature of the beam (the current inside a trajectory never changes because trajectories do not cross each other). Then we obtain

$$\frac{d^2\tau}{ds^2} = \frac{1}{4\epsilon_0}\sqrt{\frac{m}{2e}}\frac{I_\tau}{\Phi_a^{1/2}}\frac{1}{\tau(s)} \tag{10.11}$$

as the ray-trajectory equation with I_τ being constant along the (laminar) ray. This equation is easily solved, but it and the solution are more easily expressed in terms of two useful parameters,

$$K_\tau \equiv \beta\frac{I_\tau}{\Phi_a^{3/2}}, \qquad \beta \equiv \frac{1}{2\pi\epsilon_0}\sqrt{\frac{m}{e}} \approx 30.308(\mathrm{V}^{3/2}/\mathrm{mA}). \tag{10.12}$$

The *perveance* K_τ is a characteristic space-charge parameter, and it is independent of s if Φ_a is a constant (e.g., in a lens-free drift region). Because I_τ is usually less than 3.5 mA and because the ultor voltage V of the drift region in a TV kinescope is usually less than 30 kV, it follows that the beam perveance in TV kinescopes is usually less than 2×10^{-5}. Note that the perveance is a dimensionless parameter, and observe that we leave off the subscript "τ" when the entire current is implied, or when we are not interested in the contents of more than one trajectory tube. The fact that the perveance is very small is quite important in much of the following development. At any rate, Eq. (10.11) becomes

$$\frac{d^2\tau}{ds^2} = \frac{K_\tau}{2\tau}. \tag{10.13}$$

Multiplication of both sides of this equation by the integrating factor $2d\tau/ds$ yields

$$\frac{d}{ds}\left[\left(\frac{d\tau}{ds}\right)^2 - K_\tau \ln \tau(s)\right] = 0. \tag{10.14}$$

We choose as initial condition that the "waist" of the beam is at $s = 0$, i.e., that $d\tau(s)/ds = 0$ for $s = 0$. The first integration of Eq. (10.14) then yields

$$\frac{d\tau}{ds} = \sqrt{K_\tau \ln\left(\frac{\tau(s)}{\tau(0)}\right)}. \tag{10.15}$$

The next integration, to obtain $\tau(s)$, is obtained most conveniently by considering τ as the independent and s as the dependent variable. Let us abbreviate $\tau(0) = \tau_0$ for convenience, to obtain

$$s = \int_{\tau_0}^{\tau} d\tau_1 \, \frac{1}{\sqrt{K_\tau \ln(\tau_1/\tau_0)}} , \tag{10.16}$$

which is interpreted more easily by defining the new variable ξ_1 with $\xi_1^2 \equiv \ln(\tau_1/\tau_0)$ so that $\tau_1 = \tau_0 \exp(\xi_1^2)$ and $d\tau_1 = 2\tau_1 \xi_1 \, d\xi_1$:

$$\int_0^{\xi} d\xi_1 \exp(\xi_1^2) = \frac{1}{2} \sqrt{K_\tau} \, \frac{s}{\tau_0} . \tag{10.17}$$

We know that $K_\tau = \beta I_\tau / \Phi_a^{3/2}$ from Eq. (10.12), and we have $I_\tau = J(0)\pi\tau_0^2$ at $s = 0$ as the current inside the cylindrical tube obtained by rotating the trajectory $\tau(s)$ with $\tau(0) = \tau_0$ around the axis. Consequently,

$$\int_0^{\xi} d\xi_1 \exp(\xi_1^2) = \frac{1}{2} \sqrt{\frac{\pi\beta J(0)}{\Phi_0^{3/2}}} \, s . \tag{10.18}$$

The left-hand side integral is known as *Dawson's integral* (Abramowitz and Stegun 1964), and we can calculate $\xi(s)$ from it for any given value of s. Equation (10.18) has an interesting consequence: all ratios of trajectory radii at $s = 0$ are maintained as the parameter s is progressed towards greater values, i.e., the beam remains uniform if it was so at $s = 0$, and thus "swells" uniformly. That is because $\xi(s)$ is determined uniquely from the right-hand side of Eq. (10.18), and the definition $\xi^2(s) = \ln[\tau(s)/\tau_0]$ expresses the uniqueness at fixed s of the expansion factor $\tau(s)/\tau_0$ no matter which trajectory is being followed. The value $\exp[\xi^2(s)]$ is the factor by which the radius of the beam has expanded between 0 and s.

Television kinescope currents are typically 3.5 mA or less, and the ultor voltage (i.e., the voltage of the long drift region between electron gun and screen) is usually chosen to be in the 20–30 kV range. If the largest of these values are applied to Eq. (10.12) for the perveance of a beam, we obtain $K \sim 2 \times 10^{-5}$, and if Eq. (10.18) is then applied to an electron beam drifting 13.5 in to the TV screen, we find,

$$\begin{aligned} \tau(s)/\tau(0) &\approx 1.91 \text{ for a 0.030-in radius beam} \\ &\approx 1.35 \text{ for a 0.050-in radius beam} \end{aligned}$$

With the above, we have at least shown that assumption (iii) is self-consistent. When the ratio $\tau(s)/\tau(0)$ is close to unity we may employ cylindrical coordinates, set $\tau(s) \approx \rho(0)$ in the denominator of the right-hand side of Eq. (10.13), and replace the left-hand side by $d^2\rho/dz^2$ to obtain

$$\rho(z) \approx \rho(0) + \frac{K_\rho}{4\rho(0)} z^2 . \tag{10.19}$$

The beam width changes *parabolically* with z at its waist. Finally, let us check assumption (ii) to see if indeed the longitudinal space-charge force is negligible. Coulomb's law for the entire space-charge force is

$$\begin{aligned} E(\boldsymbol{r}) &= \frac{1}{4\pi\epsilon_0} \iiint d^3r_1 \, \omega(\boldsymbol{r}_1)\vec{\nabla}_1\left[\frac{1}{|\vec{r} - \vec{r}_1|}\right] \\ &= -\frac{1}{4\pi\epsilon_0} \iiint d^3r_1 \frac{1}{|\vec{r} - \vec{r}_1|} \vec{\nabla}\omega(\vec{r}_1) . \end{aligned} \tag{10.20}$$

The second integral follows from integration by parts of the first integral with a vanishing surface-integral term (because $\omega(\vec{r}_1) = 0$ as $|\vec{r}| \to \infty$). We also have seen that the charge density is related to the current density by

$$\omega(s) = \frac{I}{\pi\tau^2 v(s)} = \sqrt{\frac{2e}{m}} \frac{I}{\pi\tau^2 \Phi_a^{1/2}} , \tag{10.21}$$

from which we find, for constant or almost constant accelerating potential Φ_a, that at location s and at radius τ,

$$\frac{\partial\omega}{\partial s} \approx -\frac{2\omega}{\tau}\frac{\partial\tau}{\partial s} , \qquad \vec{\nabla}_\tau \omega \approx -\frac{2\omega}{\tau}\hat{\tau} . \tag{10.22}$$

These are the longitudinal and transverse components, respectively, of the gradient of the charge density. We observe from insertion of Eq. (10.15) for the factor $\partial\tau/\partial s$ that $|\partial\omega/\partial s| \ll |\vec{\nabla}_\tau \omega|$. In fact, we find that $|\partial\omega/\partial s| \sim K_\tau^{1/2}|\vec{\nabla}_\tau \omega|$ if we ignore the factor $\exp(\xi^2)$ because it is of order unity. As these two partial derivatives of charge density determine the longitudinal and transverse electric field strengths, respectively, according to Eq. (10.20), it follows that assumption (ii) has also been confirmed whenever the perveance is very small.

We are now in a position to check assumption (i), which we do from

$$\Phi(\vec{r}) = \frac{1}{4\pi\epsilon_0} \iiint d^3r_1 \frac{\omega(\vec{\rho}_1, z)}{|\vec{r} - \vec{r}_1|} . \tag{10.23}$$

Let ρ_0 be the effective radius of the beam and replace $\omega(\vec{\rho}_0, z)$ by $\sqrt{2e/m}I/\pi\rho_0^2\Phi_a^{1/2}$, to obtain

$$\frac{\Phi(\vec{r})}{\Phi_a} = \frac{1}{2\pi} K \iiint d^3r_1 \frac{\omega(\vec{\rho}_1, z_1)}{\omega(\vec{\rho}_0, z_1)} \frac{1}{\rho_0^2|\vec{r}-\vec{r}_1|} . \tag{10.24}$$

The first factor in the integral is a dimensionless shape factor of the beam density at z_1. For example, it might be $\exp(-\rho_1^2/\rho_0^2)$ for a Gaussian profile of charge density in a plane perpendicular to the beam direction. Under the rough assumption that the transverse integral over d^2r_1 of all but the $1/|\vec{r}-\vec{r}_1|$ factor is unity, and the additional assumption that the length of the beam is sufficient to set $|\vec{r}-\vec{r}_1| \approx |z - z_1|$, we obtain $\Phi(\vec{r})/\Phi_a \sim (1/2\pi)K \ln[L_2/L_1]$, where L_1 and L_2 are the lengths of the beam left and right, respectively, from z. This approximate calculation is sufficient to clarify that the condition $K \ll 1$ suffices to guarantee that $|\Phi(\vec{r})| \ll \Phi_a$ for the space-charge potential $\Phi(\vec{r})$.

10.2 THE LAMINAR UNIFORM PLANAR BEAM

Another situation that is easily calculated is that of a *planar* beam of electrons with all initial velocities in the z direction and determined by a single accelerating potential $\Phi_a(z)$. The calculation resembles that of Section 10.1, as might be expected. The beam is sketched in Fig. 10-2 in which two planes containing parallel trajectories in each are shown (one above and one below the $y = 0$ plane). The symmetry is specified by the condition that fields and charge densities are not a function of the x coordinate, so that the governing Maxwell equation becomes

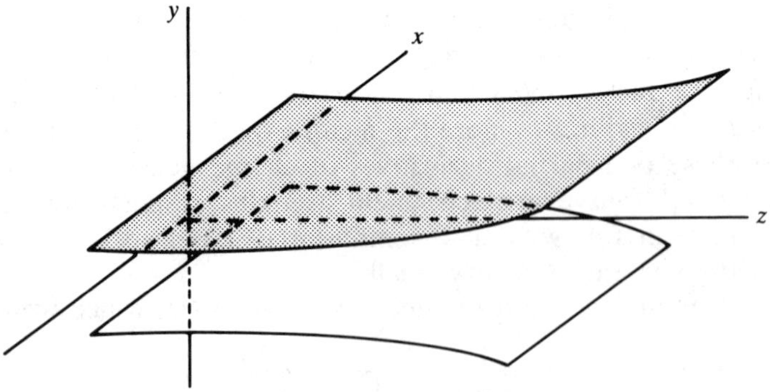

Fig. 10-2. A portion of a laminar uniform planar beam.

$$\frac{\partial E_y}{\partial y} + \frac{\partial E_z}{\partial z} = \frac{1}{\epsilon_0}\,\omega(y, z)\,. \tag{10.25}$$

The three approximations made in Section 10.1 in solving Eq. (10.7) do not depend crucially upon the geometrical symmetries and therefore they can be adapted to the present situation easily. We drop the longitudinal-field term to obtain

$$E_\tau(\tau, s) = \frac{1}{\epsilon_0}\int_0^\tau d\tau_1\,\omega(s) = -\frac{J(s)}{\epsilon_0 v(s)}\,\tau\,, \tag{10.26}$$

where s and τ are mutually orthogonal coordinates in an $x = \text{constant}$ plane, and $J(s)$ is the current density. The amount of current enclosed per unit length in the x direction between $\tau(s)$ and $-\tau(s)$ is $dI_\tau/dx = 2\tau J(s)$. In analogy to Eq. (10.10), we obtain here for the transverse force equation,

$$\frac{d^2\tau}{ds^2} = \left(\frac{e}{m\epsilon_0}\right)\frac{J(s)}{v^3(s)}\,\tau = \pi\beta\,\frac{J(s)}{\Phi_a^{3/2}}\,\tau(s) = \frac{1}{2}\,\pi\beta\,\frac{dI_\tau/dx}{\Phi_a^{3/2}}\,. \tag{10.27}$$

The right-hand side of this equation is not a function of s (under the approximations we have assumed) so that integration is straightforward:

$$\tau(s) = \tau(0) + \frac{1}{4}\,\pi\beta\,\frac{dI_\tau/dx}{\Phi_a^{3/2}}\,s^2\,, \tag{10.28}$$

with the initial condition that the beam starts at its waist, i.e., that $d\tau/ds = 0$ at $s = 0$. As in the case of the axisymmetric beam, the ratio $\tau(s)/\tau(0)$ depends only on s, and therefore an initially uniform beam remains uniform as s progresses. However, the uniform planar beam will continue to expand parabolically, in contrast to the uniform axisymmetric beam which only does so initially according to Eq. (10.19). Moreover, the axisymmetric beam diverges more rapidly as $\rho(0) \to 0$ because the space-charge fields needed to maintain constant total current will diverge. This is not so for $\tau(0) \to 0$ in Eq. (10.28), even though $J(s) \to \infty$, because the current can expand "freely" in the x direction.

10.3 A GENERAL NUMERICAL TREATMENT

When a beam has a straight-line reference trajectory, e.g., the z axis, the trajectory equation for motion of the charged particles in an electrostastic field is given by Eq. (3.19),

$$\vec{\rho}''(z) = -\frac{1+\vec{\rho}'^2(z)}{2\Phi_a}\,[\vec{E}_\rho(\vec{\rho}, z) - \vec{\rho}'(z)E_z(\vec{\rho}, z)]\,. \qquad (10.29)$$

It is not difficult to include magnetic fields in the expression, but they will be left out for the sake of clarity. The transverse electric field in this expression,

$$\vec{E}_\rho(\vec{\rho}, z) = \vec{E}_\rho^{\,\mathrm{Lp}}(\vec{\rho}, z) + \vec{E}_\rho^{\,\mathrm{sc}}(\vec{\rho}, z)\,, \qquad (10.30)$$

is the sum of a Laplace term (which is independent of the beam) and a space-charge term (which is determined solely by the beam). We have shown in Sections 10.1 and 10.2 that the longitudinal space-charge field is negligible when the perveance $K \ll 1$. Therefore, Eq. (10.29) is reasonably well approximated by

$$\vec{\rho}''(z) \approx -\frac{1+\vec{\rho}'^2(z)}{2\Phi_a}\,[\vec{E}_\rho^{\,\mathrm{Lp}}(\vec{\rho}, z) + \vec{E}_\rho^{\,\mathrm{sc}}(\vec{\rho}, z) - \vec{\rho}'(z)E_z^{\mathrm{Lp}}(\vec{\rho}, z)]\,. \qquad (10.31)$$

The space-charge electric field is given by

$$\vec{E}_\rho^{\,\mathrm{sc}}(\vec{r}) = \frac{1}{4\pi\epsilon_0}\iiint d^3r_1\, \omega(\vec{\rho}_1, z_1)\,\frac{\vec{\rho} - \vec{\rho}_1}{|\vec{r} - \vec{r}_1|^3}\,, \qquad (10.32)$$

if $\vec{r}_1 = (\vec{\rho}_1, z_1)$ and $\vec{r} = (\vec{\rho}, z)$. The beam expands slowly when perveance $K \ll 1$, and $\omega(\vec{\rho}_1, z_1)$ may be regarded as being approximately constant in z_1 around z over a range in which $|z - z_1|$ varies essentially from zero to infinity. With that approximation (for thin beams), we can perform the dz_1 integration to obtain

$$\vec{E}_\rho^{\,\mathrm{sc}}(\vec{r}) = \frac{1}{2\pi\epsilon_0}\iint d^2\rho_1\, \omega(\vec{\rho}_1, z_1)\,\frac{\vec{\rho} - \vec{\rho}_1}{|\vec{\rho} - \vec{\rho}_1|^2}\,, \qquad (10.33)$$

because

$$\int_{-\infty}^{\infty} dz_1[(\vec{\rho} - \vec{\rho}_1)^2 + (z - z_1)^2]^{-3/2} = \frac{2}{|\vec{\rho} - \vec{\rho}_1|^2}\,. \qquad (10.34)$$

It is not difficult to demonstrate that the difference between Eqs. (10.33) and (10.32) lies in the neglect of terms of $O(D|\partial\omega/\omega\partial z|)$, where D is a measure of the beam diameter at z.

The advantage of Eq. (10.33) over (10.32) lies in the consequences for solving Eq. (10.31) numerically. The iterations noted in connection with

Eqs. (10.1)–(10.4) can be avoided because we now have a pure marching method available in, for example, Eq. (10.3), repeated below:

$$\vec{\rho}(z+dz)=\vec{\rho}(z)+\vec{\rho}\,'(z)dz+\tfrac{1}{2}\vec{\rho}\,''(z)(dz)^2+\cdots\,,$$
$$\vec{\rho}\,'(z+dz)=\vec{\rho}\,'(z)+\vec{\rho}\,''(z)dz+\cdots\,. \tag{10.35}$$

If we use Eq. (10.31), with $\vec{E}^{\,\mathrm{sc}}_{\rho}$ given by Eq. (10.33), to replace $\vec{\rho}\,''(z)$ in both of these equations, then clearly all terms on the right-hand side are evaluated at z, and iteration is no longer required. Just one set of evaluations of the transverse space-charge force given in Eq. (10.33) at various $\vec{\rho}$ but fixed z suffice to obtain $\vec{\rho}(z+dz)$ from $\vec{\rho}(z)$. To work this out a little further, we again replace $\omega(\vec{\rho}_1, z)$ by $-\sqrt{m/2e}J(\vec{\rho}_1, z)/\Phi_a^{1/2}$ to obtain

$$\frac{1}{2\Phi_a^{1/2}}\,\vec{E}^{\,\mathrm{sc}}_{\rho}(\vec{r})\approx-\frac{1}{2}\,\beta\,\frac{I}{\Phi_a^{3/2}}\iint d^2\rho_1\,\frac{J(\vec{\rho}_1, z)}{I}\,\frac{\vec{\rho}-\vec{\rho}_1}{|\vec{\rho}-\vec{\rho}_1|^2}\,. \tag{10.36}$$

Here, $\beta I/\Phi_a^{3/2}=K$ is the perveance which is usually a very small quantity compared to unity. Let us define a normalized density

$$j(\vec{\rho}_1, z)\equiv\frac{1}{I}\,J(\vec{\rho}_1, z)\,, \tag{10.37}$$

so that the surface cross-sectional integral of $j(\vec{\rho}_1, z)$ is unity. We then obtain

$$\frac{1}{2\Phi_a^{1/2}}\,\vec{E}^{\,\mathrm{sc}}_{\rho}(\vec{r})\approx-\frac{1}{2}\,K\iint d^2\rho_1\;j(\vec{\rho}_1, z)\,\frac{\vec{\rho}-\vec{\rho}_1}{|\vec{\rho}-\vec{\rho}_1|^2}\,. \tag{10.38}$$

Because the factor $1+\vec{\rho}\,'^2$ may be approximated by unity not far from the waist when $K\ll 1$, it follows that Eq. (10.38) represents the space-charge field entirely in Eq. (10.31).

Although the previous equations are for beams of arbitrary shape, let us first consider axisymmetric beams for which the normalized current density $j(\rho_1, z)$ is not a function of the azimuthal angle ϕ. The angular integration in Eq. (10.38) can then be performed (it is a standard "look-up" integral) to yield,

$$\frac{1}{2\Phi_a^{1/2}}\,\vec{E}^{\,\mathrm{sc}}_{\rho}(\vec{r})\approx-K\,\frac{\pi}{\rho}\int_0^{\rho}d\rho_1\rho_1\;j(\vec{\rho}_1, z)\hat{\rho}\,. \tag{10.39}$$

The same result can also be obtained from Eqs. (10.9) and (10.10) upon equating τ with ρ and s with z, but with omission of the last step from (10.9). An especially simple case is that of a Gaussian current-density profile,

$$j(\rho_1, z) = \frac{1}{\pi\rho_0^2} \exp\left(-\frac{\rho_1^2}{\rho_0^2}\right), \tag{10.40}$$

in which the $1/e$ beam radius $\rho_0(z)$ may well be a function of z. Substitution into the space-charge field (10.39) yields

$$\frac{1}{2\Phi_a^{1/2}} E_\rho^{sc}(\vec{r}) = -\frac{1}{2} K \frac{1-\exp(-\rho^2/\rho_0^2)}{\rho}, \tag{10.41}$$

which has a maximal value at $\rho \approx 1.12\rho_0$. It is easily seen that beam profiles with maximum value of $j(\rho_1, z)$ on the axis, that fall off monotonically with increasing ρ_1, have asymptotic forms similar to that of Eq. (10.41) for $\rho \to 0$ and $\rho \to \infty$. For $\rho \to 0$ we may approximate Eq. (10.39) by

$$\frac{1}{2\Phi_a^{1/2}} \mathrm{E}_\rho^{sc}(\vec{r}) \approx -\frac{1}{2} K\pi j(0, z)\rho = -\frac{1}{2} K\pi\rho \frac{J(0, z)}{I} \approx -\frac{1}{2} K \frac{\rho}{\rho_0^2}, \tag{10.42a}$$

if we define the effective radius at z by $\rho_0^2 \equiv I/\pi J(0, z)$. In the other limit, $\rho \to \infty$, we may set the upper bound of the integral in Eq. (10.39) equal to infinity to obtain

$$\frac{1}{2\Phi_a^{1/2}} E_\rho^{sc}(\vec{r}) \approx -\frac{1}{2} \frac{K}{\rho}. \tag{10.42b}$$

Hence such beams always have a transverse space-charge field profile as sketched in Fig. 10-3.

This information inspires the following more general treatment of beams which are not necessarily axisymmetric. First, we discretize Eq. (10.38),

$$\frac{1}{2\Phi_a^{1/2}} \vec{E}_\rho^{sc}(\vec{r}) \approx -\frac{1}{2} K \sum_m \sum_n j_{mm} \vec{L}^{mn}(\vec{\rho}),$$

$$j_{mn} \approx j(x_m, y_n), \tag{10.43}$$

$$\vec{L}^{mn}(\vec{\rho}) = \int_{x_m - \frac{1}{2}h}^{x_m + \frac{1}{2}h} d\xi \int_{y_n - \frac{1}{2}h}^{y_n + \frac{1}{2}h} d\eta \frac{(x-\xi)\hat{x} + (y-\eta)\hat{y}}{(x-\xi)^2 + (y-\eta)^2}.$$

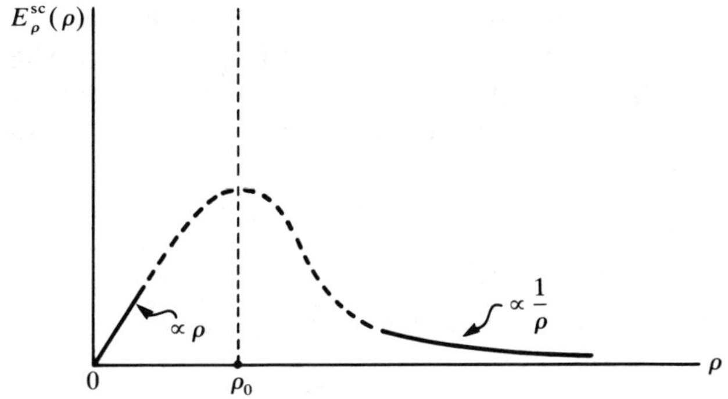

Fig. 10-3. Profile of the transverse space-charge field in a beam with monotonically decreasing axial density.

The discretization of the cross-section of the beam is into a two-dimensional mesh of squares with $x_m = mh$ and $y = nh$ ($-\infty < m, n < \infty$), and with the origin of the integers m, n chosen to be on a central trajectory (see Fig. 10-3). We can write $\vec{L}^{mn}(\vec{\rho}) = L_x^{mn}(\vec{\rho})\hat{x} + L_y^{mn}(\vec{\rho})\hat{y}$, with obvious expressions for the two parts derived from the definition in Eq. (10.43). It will be sufficient to discuss $L_x^{mn}(\vec{\rho})$ in more detail; the corresponding results for $L_y^{mn}(\vec{\rho})$ can be found afterwards by exchanging the roles of the x, ξ and y, η variables with each other. Straightforward integration of the double integral for $L_x^{mn}(\vec{\rho})$ in Eq. (10.43) results in an analytical expression that we write as follows:

$$L_x^{mn}(\vec{\rho}) = \frac{1}{2}(\Delta y_+)\ln\left(\frac{\Delta x_+^2 + \Delta y_+^2}{\Delta x_+^2 + \Delta y_-^2}\right) - \frac{1}{2}(\Delta y_-)\ln\left(\frac{\Delta x_-^2 + \Delta y_+^2}{\Delta x_-^2 + \Delta y_-^2}\right)$$
$$+ 2(\Delta x_+)\left[A\left(\frac{\Delta y_+}{\Delta x_+}\right) - A\left(\frac{\Delta y_-}{\Delta x_+}\right)\right]$$
$$- 2(\Delta x_-)\left[A\left(\frac{\Delta y_+}{\Delta x_-}\right) - A\left(\frac{\Delta y_-}{\Delta x_-}\right)\right], \tag{10.44a}$$

where $A(x)$ is an abbreviation for the arctangent of x, and where the following definitions apply:

$$\begin{aligned} \Delta x_+ &\equiv x - x_m + \tfrac{1}{2}h\,, \qquad & \Delta y_+ &\equiv y - y_n + \tfrac{1}{2}h\,, \\ \Delta x_- &\equiv x - x_m - \tfrac{1}{2}h\,, \qquad & \Delta y_- &\equiv y - y_n - \tfrac{1}{2}h\,. \end{aligned} \tag{10.44b}$$

The evaluation of L_x^{mn} and L_y^{mn} for given $\vec{\rho}$ and one pair of values of m, n

requires evaluation of at least four arctangents and four natural logarithms. This could imply a costly numerical procedure when there are many mesh squares inside the beam contour (i.e., in regions where j_{mn} is not negligibly small). However, the following asymptotic forms follow from Eqs. (10.44):

$$\begin{aligned} L_x^{mn} &\approx \frac{x - x_m}{|\Delta\vec{\rho}_{mn}|^2} \quad \text{for } |x - x_m| \gg \frac{1}{2}\, h\,, \\ L_x^{mn} &\approx \pi(x - x_m)/h^2 \quad \text{for } |x - x_m| \ll \tfrac{1}{2} h\,, \end{aligned} \tag{10.45}$$

where $\Delta\vec{\rho}_{mn} \equiv (x - x_m)\hat{x} + (y - y_n)\hat{y}$. Because these asymptotic forms hold outside the hatched region sketched in Fig. 10-4 (for given $\vec{\rho}$), and because the first of the asymptotic forms (10.45) is a good approximation over all but a small part of the beam close to $\vec{\rho}$ if h is sufficiently small, it follows that an interpolation formula is useful:

$$L^{mn}(\vec{\rho}) \approx \frac{\Delta\vec{\rho}_{mn}}{|\Delta\vec{\rho}_{mn}|^2 + (h^2/\pi)}\,, \tag{10.46}$$

which reduces to (10.45) when $|\Delta\vec{\rho}_{mn}|$ is either much larger or smaller than $\frac{1}{2}h$. Because the region $|\Delta\vec{\rho}_{mn}| \ll \frac{1}{2}h$ is so small, one might wonder why the term h^2/π is included in the denominator of Eq. (10.46). The reason is that this extra term corrects properly for the singularity that might otherwise occur in the mathematics (but not in the physics of the problem) when the force very close to the point (x_m, y_n) is being evaluated. Physically, we know that the Coulomb forces tend to zero in the immediate vicinity of an interior point (Kirstein et al. 1967).

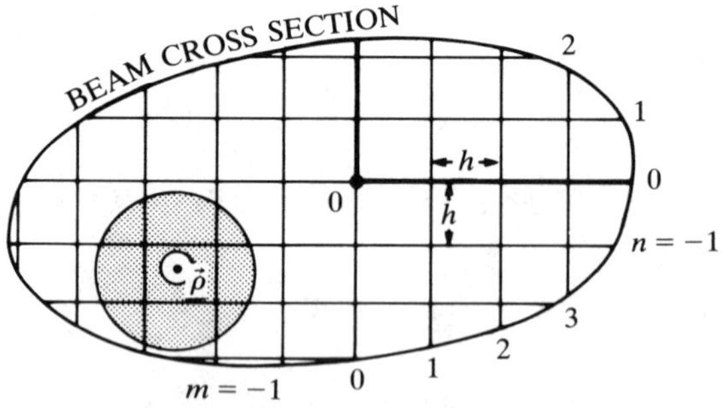

Fig. 10-4. Illustration of the regions in which a simplified space-charge field approximation does (open) and does not (shaded) hold.

The result of the above then yields

$$\frac{1}{2\Phi_a^{1/2}} \vec{E}_\rho^{\,\mathrm{sc}}(\rho, \vec{z}) \approx -\frac{1}{2} K \sum_m \sum_n j_{mn}(z) \frac{\Delta\vec{\rho}_{mn}}{|\Delta\vec{\rho}_{mn}|^2 + (h^2/\pi)} , \quad \Delta\vec{\rho}_{mn} \equiv (x - x_m)\hat{x} + (y - y_n)\hat{y} . \tag{10.47}$$

Various simulations of electron guns (de Wolf 1982) have shown that this approximation, even with relatively large mesh parameter h, yields highly acceptable results for trajectory calculations which include space-charge effects.

10.4 THE SIMPLEST DIODE EQUATION

Consider the planar diode in which a planar cathode at $z = 0$ emits electrons, essentially at zero velocity. If the anode is a plane at $z = d$ (see Fig. 10-5) with voltage difference V_0 (with respect to the cathode voltage) and with negligible space charge, then Laplace's equation will predict trivially an electric field $\vec{E} = -(V_0/d)\hat{z}$ between the cathode and anode planes. This is sketched in Fig. 10-6. However, a more precise description should take into account a "shielding" effect when electrons leave the cathode. These electrons exert a repulsive force on electrons about to leave the cathode; this expresses itself as a reduction of the Laplace-induced electric field $E_z^{\mathrm{Lp}} = -V_0/d$. The electric field at $z = 0$ can actually be driven back to zero if the cathode can emit enough electrons. A dynamically stable situation arises in which $E_z = 0$ and which produces a steady flow of current to the anode as long as the cathode can produce electrons. This is described in the literature as *space-charge-limited*

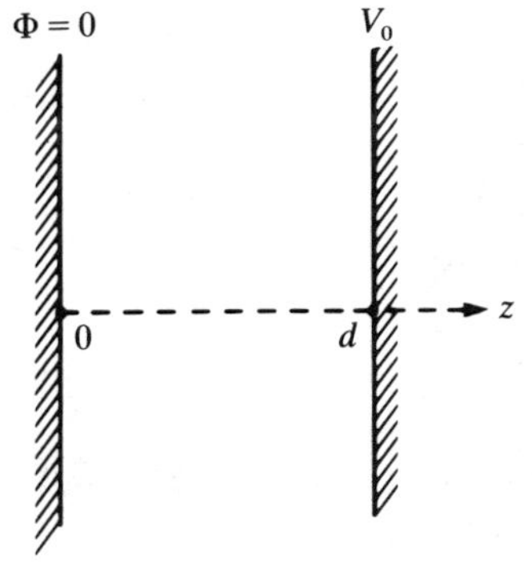

Fig. 10-5. A planar diode.

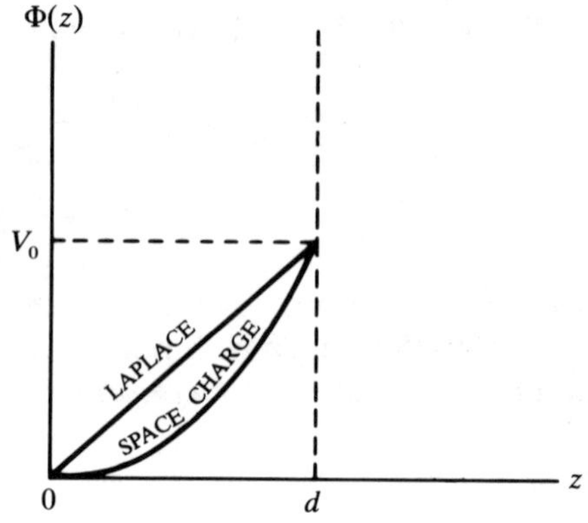

Fig. 10-6. The potential in a planar diode with and without space charge.

current. The situation is even more complex when it is looked into in more detail. The electrons leaving a cathode at temperature T have a half-Maxwellian distribution of kinetic energies, $\frac{1}{2}mv_z^2$ distributed around the energy $\frac{1}{2}kT$ (Kirstein et al. 1967). That allows E_z to become somewhat negative at $z = 0$, and the location where E_z is actually zero for space-charge limited flow lies slightly in front of the cathode (and is regarded as the *virtual* cathode). Because eV_0 is usually very large in comparison to kT, we may nearly always ignore this small effect of a slightly displaced virtual cathode at all locations except immediately in front of the cathode.

A quantitative description now follows, under the assumption that the business about a virtual cathode may be ignored, starting from Poisson's equation:

$$\frac{d^2\Phi}{dz^2} = -\frac{1}{\epsilon_0}\,\omega(z) = 2\pi\beta\,\frac{J_s(z)}{\Phi^{1/2}}\,. \tag{10.48}$$

We again have used the fact that $\omega(z) = -J_s(z)/v(z)$, and that the initial kinetic energy of the electrons can be ignored in comparison to $\frac{1}{2}mv^2(z)$ everywhere except very close to the cathode. The constant β was defined previously in Eq. (10.12) and it appears when we replace $v(z)$ by $(2e\Phi/m)^{1/2}$. The solution to Eq. (10.48) is obtained by means of the multiplying factor $d\Phi/dz$ which modifies the equation into

$$\frac{d}{dz}\,((\Phi')^2 - 8\pi\beta J_s\Phi^{1/2}) = 0\,. \tag{10.49}$$

The space-charge limited solution is obtained by using the initial condition $\Phi'(0) = 0$ (we already have $\Phi = 0$ at the cathode, $z = 0$):

$$\Phi'(z) = \sqrt{8\pi\beta J_s(z)}\Phi^{1/4} \quad \text{or} \quad \frac{d}{dz}\,[\Phi^{3/4}(z)] = \frac{3}{4}\sqrt{8\pi\beta J_s(z)}\,. \qquad (10.50)$$

With the initial condition of zero potential at $z = 0$, this integrates to

$$\Phi(z) = \left(\frac{9\pi\beta}{2}\,J_s\right)^{2/3} z^{4/3}\,. \qquad (10.51)$$

This equation and a number of variations for other geometries date from very early work in electron optics, reviewed authoritatively by Ivey (1954). Note that J_s in Eq. (10.51) cannot be a function of z: the current flows in the z direction and is stationary, which implies from the equation of continuity that $dJ_s/dz = 0$. The magnitude of J_s is obtained by applying Eq. (10.51) to the anode at $z = d$ where $\Phi(d) = V_0$:

$$J_s = \frac{2}{9\pi\beta}\,\frac{V_0^{3/2}}{d^2}\,, \qquad (10.52)$$

known as the *Child-Langmuir* law. It predicts that J_s is proportional to the three-halves power of the anode-cathode voltage difference. Insertion of Eq. (10.52) into (10.51) yields the simple relationships,

$$\Phi(z) = V_0\left(\frac{z}{d}\right)^{4/3}\,, \qquad E(z) = \frac{4}{3}\,E_z^{\mathrm{Lp}}\left(\frac{z}{d}\right)^{1/3}\,, \qquad (10.53)$$

given that $E_z^{\mathrm{Lp}} = -V_0/d$. The four-thirds dependence of the space-charge limited diode potential $\Phi(z)$ upon z is also depicted in Fig. 10-6. The reader is referred once again to Ivey (1954) for similar results in cylindrical and spherical diodes.

Limited space-charge potentials and fields hold in those cases that the anode does not exert enough force upon the cathode to extract a large enough cloud of electrons that will drive E_z to zero. In the case of limited space-charge flow, the derivative $\Phi'(0)$ will have a value between 0 and V_0/d. Instead of Eq. (10.50), we then obtain

$$\Phi'(z) = \sqrt{8\pi\beta J_s(z)\Phi^{1/2}(z) + [\Phi'(0)]^2}\,. \qquad (10.54)$$

Let us rename $\Phi'(0) \equiv -E_0$, and $V_0/d = -E_0^{\mathrm{Lp}}$. Let J_s be the saturation current density given by Eq. (10.52) for space-charge limited flow. We expect to obtain $0 < J < J_s$. Let us also define the normalized field $e \equiv \frac{3}{4}E_0/E_0^{\mathrm{Lp}}$, the normalized potential $v \equiv \Phi/V_0$, the normalized current density $j \equiv J/J_s$, and the normalized distance $\zeta \equiv z/d$. Integration of Eq. (10.54) with the initial condition $\Phi(0) = 0$ will then yield

$$(jv + e^2)^{3/2} - 3e^2(jv + e^2) = j^2\zeta - 2e^3 \,. \qquad (10.55)$$

The new parameters are subject to the limits $0 < e < \frac{3}{4}$, $0 < \zeta < 1$, $0 < v < 1$, and $0 < j < 1$. Simple algebraic manipulation of Eq. (10.55) results in a quadratic equation in j:

$$\zeta^2 j^2 - v^3 j - e^2(4\zeta e - 3v^2) = 0 \,. \qquad (10.56)$$

To solve for v in terms of j (for given fixed values of ζ and e, i.e., for given location and initial field strength), we need to obtain j as a function of e, which we do by applying Eq. (10.56) at the anode where $\zeta = 1$, $v = 1$. We then obtain

$$\begin{aligned} j &= \tfrac{1}{2} + \tfrac{1}{2}\sqrt{1 - 12e^2 + 16e^4} \quad \text{for } 0 < e < \tfrac{1}{2} \,, \\ j &= \tfrac{1}{2} - \tfrac{1}{2}\sqrt{1 - 12e^2 + 16e^4} \quad \text{for } \tfrac{1}{2} < e < \tfrac{3}{4} \,. \end{aligned} \qquad (10.57)$$

When the dimensionless quantities are translated back into their original counterparts, this expression becomes

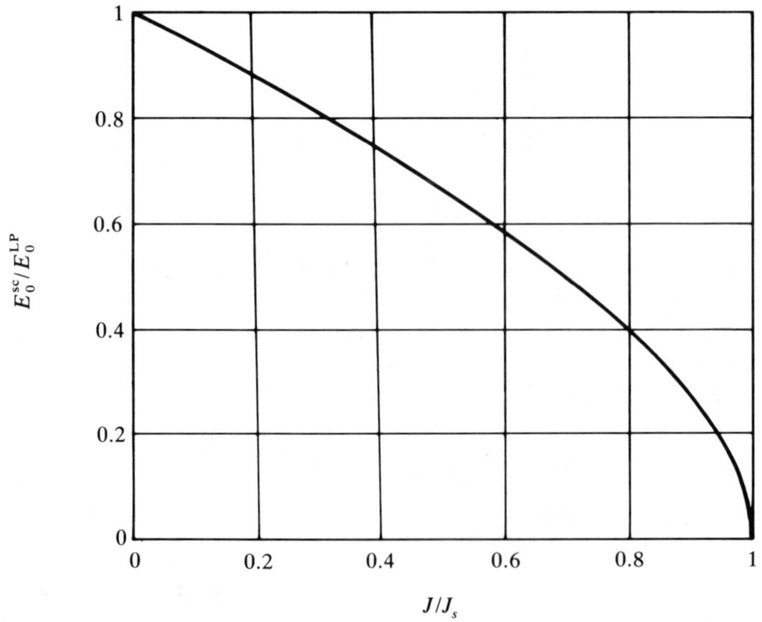

Fig. 10-7. The electric field at the cathode of a planar diode when the space-charge current is not saturated: E_0/E_0^{Lp} vs J/J_s.

$$J = \frac{1}{2} J_s\left[1 \pm \sqrt{1 - \frac{27}{4}\left(\frac{E_0}{E_0^{\mathrm{Lp}}}\right)^2\left(1 - \frac{E_0}{E_0^{\mathrm{Lp}}}\right)}\right], \tag{10.58}$$

where the upper sign holds for $|E_0| < \frac{2}{3}|E_0^{\mathrm{Lp}}|$, and the lower sign for $|E_0| > \frac{2}{3}|E_0^{\mathrm{Lp}}|$, in the understanding that $|E_0|$ is not to exceed the Laplace value $|E_0^{\mathrm{Lp}}|$. A graph of the ratio E_0/E_0^{Lp} vs J/J_s is shown in Fig. 10-7. With Eq. (10.58), we can express $\Phi(z)/V_0$ as a function of z/d and E_0/E_0^{Lp}, using Eq. (10.56), but this will not be done explicitly here.

10.5 THE PLANAR DIODE WITH RADIAL VARIATION OF ANODE VOLTAGE

A case of more than academic interest is that in which the cathode is limited electrostatically in width. In TV kinescope electron guns, lens apertures produce equipotentials that effectively limit the emitting area of the cathode. The situation is modeled best by the axisymmetric planar diode with radially varying anode voltage, $V_0(\rho, d)$ as sketched in Fig. 10-8. The potential can be expanded in terms of derivatives of the axial potential, as given in Eq. (3.70):

$$\Phi(\rho, z) = \Phi(0, z) - \tfrac{1}{4}\Phi''(0, z)\rho^2 + \tfrac{1}{24}\Phi^{(4)}(0, z)\rho^4 \cdots. \tag{10.59}$$

We will model this by means of a compact approximation:

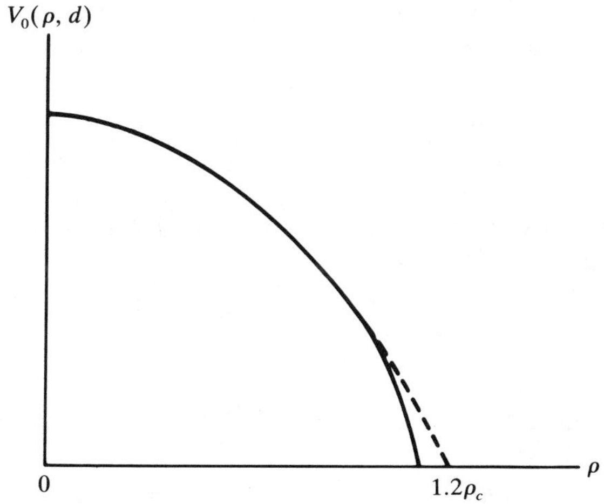

Fig. 10-8. Radial variation of anode voltage in a planar diode with radial variation due to a lens producing an effective anode. See Eqs. (10.59)–(10.60).

$$\Phi(\rho, z) = \Phi(0, z) J_0\left(\frac{2\rho}{\rho_c}\right), \qquad \rho_c^{-2} \equiv \frac{\Phi''(0, z)}{4\Phi(0, z)}, \tag{10.60}$$

where $J_0(x)$ is the Bessel function of the first kind of zero order that is finite at $x = 0$ (Abramowitz and Stegun 1964). The first zero of Eq. (10.60) occurs at $\rho \approx 1.202\rho_c$. Obviously, the expansion of (10.60) in powers of ρ will differ from Eq. (10.59) in the ρ^4 term, and in fact the error in using Eq. (10.60) for (10.59) is

$$\left[1 - \frac{\Phi^2 \Phi^{(4)}}{(\Phi'')^2}\right]\left(\frac{2\rho}{\rho_c}\right)^4 + \mathrm{O}\left(\frac{2\rho}{\rho_c}\right)^6. \tag{10.61}$$

This error would be small, more or less automatically, for $\rho \leqslant \rho_c$ if the actual value of ρ for which $\Phi(\rho, z) = 0$ lies at about 80% of $\rho = 1.202\rho_c$, and if the higher-order derivatives of Φ decrease monotonically. Fortunately, this is often the case for electronic apertures, as we shall discuss further in Chapter 11. Therefore, we shall assume that $\Phi(\rho, z)$ in a planar diode with electrostatically apertured anode voltage $V_0(\rho, d)$ is approximated well by Eq. (10.60) for $\rho \leqslant \rho_c$.

Let us rewrite Poisson's equation for axial symmetry in cylindrical coordinates:

$$\frac{\partial^2 \Phi}{\partial z^2} + \frac{1}{\rho}\left(\rho \frac{\partial \Phi}{\partial \rho}\right) = 2\pi\beta \frac{J(\rho, z)}{\Phi^{1/2}}. \tag{10.62}$$

Now use the approximation (10.60) in the form $\Phi(\rho, z) = \Phi_0(z) J_0(2\rho/\rho_c)$, where $\Phi_0(z)$ is the axial potential $\Phi(0, z)$. The properties of the J_0 Bessel function then result after the above substitution into (10.62) in

$$\Phi_0'' - 4\rho_c^{-2}\Phi_0 = 2\pi\beta \frac{J(\rho, z)}{[J_0(2\rho/\rho_c)]^{3/2}} \Phi_0^{-1/2}. \tag{10.63}$$

As before, we expect that $J(\rho, z)$ is only a function of ρ, hence write it as $J(\rho)$, and we hypothesize that

$$J(\rho) = J_c\left[J_0\left(\frac{2\rho}{\rho_c}\right)\right]^{3/2}, \tag{10.64}$$

where J_c is a constant yet to be determined. Substitution into Eq. (10.63) yields

$$\Phi_0'' - 4\rho_c^{-2}\Phi_0 - 2\pi\beta J_c \Phi_0^{-1/2} = 0, \tag{10.65}$$

which equation can be multiplied by $d\Phi_0(z)/dz \equiv \Phi_0'$ to become

$$\frac{d}{dz}\,[(\Phi_0')^2 - 4\rho_c^{-2}\Phi_0^2 - 8\pi\beta J_c\Phi_0^{1/2}] = 0 \tag{10.66}$$

This equation can be integrated with the space-charge limited flow initial conditions, $\Phi_0(0) = 0$ and $\Phi_0' = 0$, to obtain

$$\frac{d}{dz}\,(\Phi_0^{3/4}) = \left(\frac{9\pi\beta}{2}\,J_c\right)^{1/2}\left[1 + \frac{9}{4\rho_c^2}\,\frac{\Phi_0^{3/2}}{(9\pi\beta J_c/2)}\right]^{1/2}. \tag{10.67}$$

As there are some obvious combinations of parameters in here, it will be helpful to define the auxiliary parameter and variable,

$$\bar{\rho}_c \equiv \frac{2}{3}\,\rho_c\,, \qquad \xi(z) \equiv \left(\frac{9\pi\beta}{2}\,\frac{J_c}{\Phi_0^{3/2}}\right)^{-1/2}, \tag{10.68}$$

so that Eq. (10.67) takes on the simple form,

$$\xi'(z) = \left(1 + \frac{\xi^2}{\bar{\rho}_c^2}\right)^{1/2}, \tag{10.69}$$

which is easily integrated to yield a hyperbolic sine function,

$$\xi(z) = \bar{\rho}_c \sinh\left(\frac{z}{\bar{\rho}_c}\right). \tag{10.70}$$

Substitution of the definition of $\xi(z)$ into this equation yields

$$\Phi_0(z) \equiv \left(\frac{9\pi\beta}{2}\,J_c\right)^{2/3}\bar{\rho}_c^{-4/3}\left[\sinh\left(\frac{z}{\bar{\rho}_c}\right)\right]^{4/3}. \tag{10.71}$$

We obtain the desired expression for J_c by application of this equation to $z = d$, in which case we obtain the pair of equations,

$$J_c = \left[\frac{2V_0^{3/2}}{9\pi\beta d^2}\right]\left[\frac{\sinh(d/\bar{\rho}_c)}{d/\bar{\rho}_c}\right]^{-2}, \qquad \frac{\Phi_0(z)}{V_0} = \left[\frac{\sinh(z/\bar{\rho}_c)}{\sinh(d/\bar{\rho}_c)}\right]^{4/3}. \tag{10.72}$$

For the sake of completeness, we will insert these into the definition (10.64), to obtain

$$\begin{aligned} J(\rho) &= J_s\left[\frac{\sinh(d/\bar{\rho}_c)}{d/\bar{\rho}_c}\right]^{-2}\left[J_0\left(\frac{2\rho}{\rho_c}\right)\right]^{3/2}, \\ \Phi(\rho, z) &= V_0\left[\frac{\sinh(z/\bar{\rho}_c)}{\sinh(d/\bar{\rho}_c)}\right]^{4/3} J_0\left(\frac{2\rho}{\rho_c}\right). \end{aligned} \tag{10.73}$$

Here, J_s is the saturation current density of the uniform planar diode given by Eq. (10.52), and we repeat that $\bar{\rho} = \frac{2}{3}\rho_c$. The first of these is the extension of the Child-Langmuir expression for space-charge limited current in an axisymmetric planar diode, and the second gives the dependence of the potential (and field) upon axial distance z into the diode. These formulas reduce to expressions suggested on a more or less ad hoc basis by Ploke (1951) when $d \ll \bar{\rho}_c$ and $2\rho \ll \rho_c$. The new expressions do require $2\rho/\rho_c$ to be less than the first zero of the J_0 Bessel function because otherwise Eq. (10.60) is no longer a valid approximation of the potential $\Phi(\rho, z)$. The uniform planar diode expressions follow in the limit $\rho_c \rightarrow \infty$.

Applications of the above diode calculations to cathode-ray tubes (CRTs) are discussed in Chapter 11. At this stage, we close the discussion of theory and principles of electron optics, and from here on we will discuss some of the applications to the design and functioning of CRTs.

REFERENCES

Abramowitz, M. and I. A. Stegun (1964) *Handbook of Mathematical Functions*, National Bureau of Standards Appl. Math. Series 55, U.S. Department of Commerce.

de Wolf, D. A. (1982) *RCA Rev.*, **43**, 339.

Ivey, H. F (1954) Space charge limited currents, in *Advances in Electronics and Electron Physics* (Ed. L. Marton) **VI**, Academic Press, New York.

James, M. L., G. M. Smith and J. C. Wolford (1985) *Applied Numerical Methods for Digital Computations with FORTRAN and CSMP*, 3rd edn., Harper & Row, New York.

Kirtstein, P. T., G. S. Kino and W. E. Waters (1967) *Space Charge Flow*, McGraw-Hill, New York.

Ploke, M. (1951) *Z. Angew. Phys.* **3**, 441.

CHAPTER 11

ELECTRON GUNS IN TELEVISION AND OTHER CATHODE-RAY TUBES

11.1 INTRODUCTION

We are now in a position to discuss one of the major electron optical devices in use: the cathode-ray (CRT). Cathode-ray tubes play a major role in television and computer monitors. The almost universal availability of television sets and the recent growth of the personal computer industry have vastly increased the importance of the CRT, the history of which has been summarized by Grivet (1972). In television, this tube is also known as a *kinescope* (which is highly specialized towards the production of a television-screen image at relatively high data rates). The present-day interest in "high-definition television" and in improved computer monitors for graphics and other applications has also stimulated further development of the CRT. Indeed, that has been the impetus towards the writing of this text; the need to provide designers as well as researchers and students new to the field of electron optics with a lucid entry into the essentials of this discipline.

Cathode-ray tubes may differ somewhat in detail from application to application, but those that are used for example in TV kinescopes have the following common electron optics elements: an *electron gun* that forms and focusses a beam; a *drift region* that brings the beam to a spot at the desired display location (e.g., on a screen of phosphor spots or stripes); and a *magnetic or electrostatic deflector* (*yoke*) that deflects the electron beam away from the center of the display screen. These three

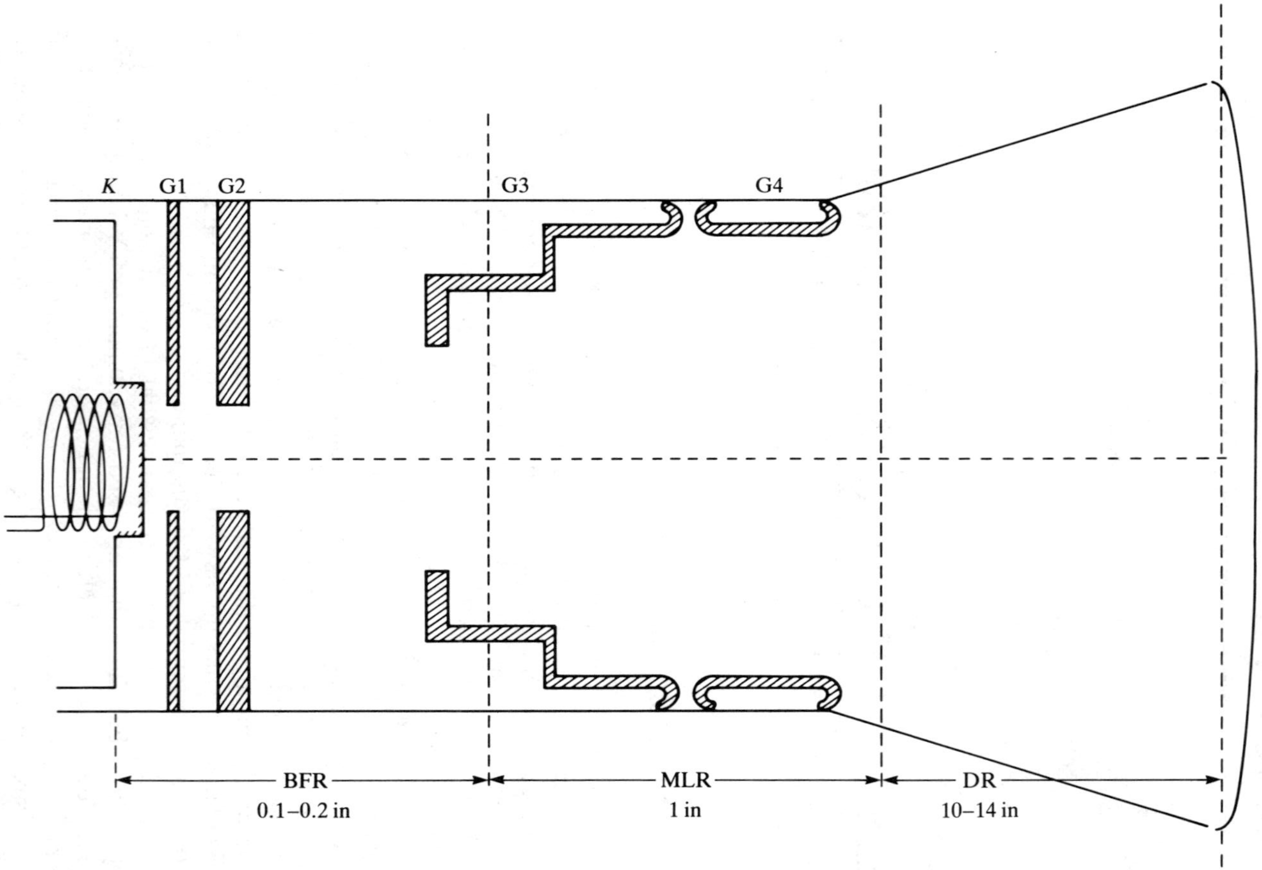

Fig. 11-1. Schematic drawing of a TV kinescope with enlargement of the electron-gun portion, and illustration of the three regions, BFR, MLR, DR.

elements are fundamentally different and, in some sense, independent of each other. One cannot speak of complete independence because each element is tailored in an overall fashion to the geometry of the others, but many details of each can be designed without much consideration of the other elements. We will not say anything further about deflection, electrostatic or magnetostatic in this section although the importance of yoke design should not be underestimated. However, as of the present, no clear principles seem to stand out, and the design of magnetic yokes, for example, remains largely an art consisting of trial and error with computer simulations that link current windings, strengths, and geometries to magnetic deflections. At any rate, the principles involved do not differ essentially from those discussed in Section 4.2.

The undeflected beam is the subject of this section; design of an optimal spot at the center of the display screen requires us then to distinguish three important regions of the CRT (see Fig. 11-1).

(I) A *beamforming region* (BFR) consisting of a cathode and several electron optics lenses which consist of closely spaced metallic plates with central (usually circular) apertures and possible also one or more cylinder lenses. The purpose of the BFR is to extract electrons from the cathode, to bring these to an early focus, and then to produce a divergent beam (which has advantages for a second focussing into a small spot).

(ii) A *main lens region* (MLR) in which a small number of cylinder lenses, usually co-linear and of the same radius held at considerably higher voltages, focus the entering divergent beam towards the display screen.

(iii) A *drift region* (DR) in which the redirected electrons move without further accelerating forces (i.e., drift) towards a spot on the screen. The length of the DR is usually an order of magnitude larger than that of the combined BFR and MLR.

The details may vary widely from one design of a kinescope or similar CRT to another, but the above separation into three regions is usually useful from a conceptual point of view.

11.2 THE BEAMFORMING REGION (BFR)

The beamforming region (BFR) is the first part of the electron gun depicted in Fig. 11-1. An informative way to understand the BFR is to consider the cathode itself first as a *thermionic* emitter of electrons. It usually consists of a block of oxide heated by coils to a temperature of the order of 1000°K. Typical oxides are those of cesium, calcium, and barium, which lower the *work function* of the metal onto which they are deposited. The work function is a quantum-mechanical surface barrier potential which electrons must overcome (or tunnel through) to escape from the cathode. The classical description is to state that the fixed positive-ion lattice tends to pull back electrons that go beyond the surface layer. We will discuss more details below.

The current density for electrons emitted at the cathode with velocities between $\vec{v} = (v_x, v_y, v_z)$ and $\vec{v} + d\vec{v} = (v_x + dv_x, v_y + dv_y, v_z + dv_z)$ is given by the Maxwell-Boltzmann distribution of kinetic gas theory (Mayer and Mayer 1966),

$$dJ_0 = \kappa \exp\left(-\frac{mv^2}{2k_B T}\right) v_z \, d^3v \,, \tag{11.1}$$

where κ is a constant, T is the temperature in degrees Kelvin (°K), and $k_B \approx 1.38 \times 10^{-23}$ J/°K is the Boltzmann constant. In the case of *thermionic emission* (which is independent of anode conditions), the surface barrier expresses itself in the fact that the constant κ is proportional to $\exp(-e\Phi_w / k_B T)$, with $e/k_B \approx 11{,}594$ in MKS units if T is in °K and e in Coulombs (C). The potential Φ_w, usually several volts in magnitude, is the above-mentioned work function representing the energy barrier (when multiplied by the electron charge e) which electrons must overcome when they are in the ground state (Kittel 1968). In the case of *space-charge limited* emission (see Section 10.4), the work function is responsible for the virtual-cathode displacement of the location where the field $E_z = 0$.

When the dependence upon work function is taken into account, integration of Eq. (11.1) over the half space $z > 0$ yields *Richardson's equation* for thermionic current (Kirstein et al. 1967):

$$J_0 = \tilde{A} T^2 \exp\left(-\frac{e\Phi_w}{k_B T}\right) . \tag{11.2}$$

The parameter $\tilde{A}$ depends upon the cathode material, but for many commonly chosen materials it is of the order of 1.2×10^{-6} A/(m °K)^2.

The effect of temperature T upon thermionic emission is quite obvious; for that reason cathode materials are heated as much as possible and as would be consistent with avoidance of damage and considerations of lifetime. *Field-emitting cathodes* are able to avoid the high-temperature problem by producing high fields locally around spike-like protuberances (thus lowering the effect of the work function) but "cold" field emission does not appear to be viable to date for electron guns. Thermionic cathodes are typically heated to a temperaturae of $T \sim 1000°K$ which corresponds to a thermal energy of $k_B T \sim 0.1$ eV so that Eq. (11.1) predicts a thermal spread of electron energies of the order of several tenths of an electron volt (eV) before the contribution to J_0 becomes negligibly small.

Hence the inescapable fact for thermionic cathode emission is that the electrons are emitted in a spectrum of energies and a spread of angles. Consequently, a lens is needed in the beamforming region (BFR) to form a beam. Thus, a "first anode" was introduced in the BFR; usually in the form of an aperture lens, i.e., a metallic plate held at a positive potential with respect to the cathode and with a more or less circular aperture to allow accelerated electrons to pass through. This lens is commonly known as the *G2 grid*. This gives rise to the situation of Fig. 11-2 which shows electrons emitted in various directions from each location with varying energies up to several times 0.1 eV at most. Many will indeed pass through the aperture. Some will terminate on the G2 grid with a circular aperture.

There are two problems with the lens action of Fig. 11-2 aside from that caused by electrons that strike the G2 grid itself. First, the actual object area is almost the entire surface of the cathode facing the interior

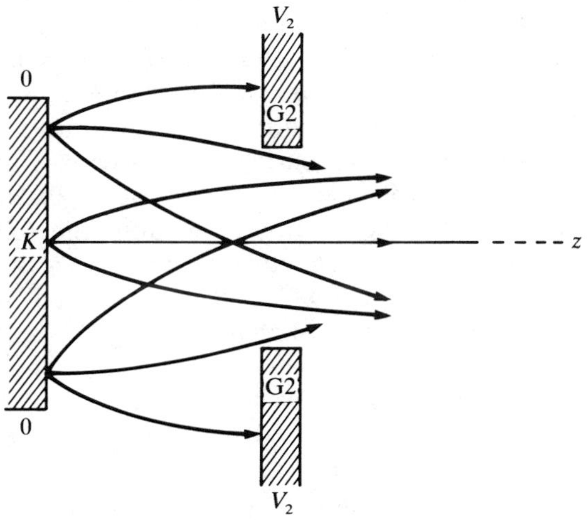

Fig. 11-2. Electron trajectories launched from a hot cathode with partial focussing by a solitary aperture lens.

of the BFR, which is fairly wide. Whatever the magnification factor may be, the resulting image according to paraxial optics (see Chapter 6) is proportional to the object area, and prohibitively large anode voltages may be needed to compensate for large object area (see Abbe's sine law, Eq. (6.8), for example) and to obtain acceptably small screen spots. In the second place, but of more direct significance, it is very difficult to *modulate* the beam current, i.e., to change its strength rapidly and accurately. To do so with only one anode would require a more or less mechanical modification of the cathode emitting area which is slow, awkward, and difficult to do. The time required for a significant change in current strength needs to be very short. To estimate it, consider that a TV screen contains perhaps 15,000 phosphor dots or so that each must be illuminated 30–60 times a second. Thus time changes are in the order of several microseconds (μs) or even less if considerations of color and chrominance, etc., are also taken into account.

The acceptable solution to this problem was found fairly rapidly in the early development; it is the insertion of a *modulator grid*, named the G1 grid, between the cathode and the G2 grid. The modulator grid is similar in appearance to the G2, but it is given a negative voltage, $-V_1$, with respect to the cathode. Thus the zero-volt equipotential will be somewhere between the G1 and the G2 and parallel to these grids when the location is sufficiently far from the axis. As one follows it down towards the axis, the contour that intersects a plane through the axis will bend towards the cathode and intersect it at a radius ρ_c which we will discuss in more detail below. The situation, with some typical equipotential contours for values close to 0 V, is shown in Fig. 11-3. Of course the zero-potential contour includes the cathode area with $0 \leqslant \rho \leqslant \rho_c$. As $-V_1$ is made more negative the value of ρ_c will decrease until at some $V_1 = V_{c0}$ the value of ρ_c has become zero and the cathode can no longer emit electrons because the field at the surface everywhere is repulsive. It is customary to design the BFR such that $V_{c0} = 150$ V. Thus, the G1 grid acts as an *electronic shutter* operated by modulating the voltage $-V_1$ and so increasing or decreasing the effective emitting area of the cathode. There is also a second advantage which can be deduced by observing a number of electron trajectories as sketched in Fig. 11-3. Principal electrons will focus at z_1 in the absence of aberrations and space-charge effects which actually cause the focus to "smear" over a nonzero axial distance around z_1. Electrons emitted at angles close to grazing will intersect the principal electrons at locations such as z_2 and z_2'. There will be a minimal beam size somewhere between these latter two locations. The distribution of current density will tend to weight it towards the vicinity of z_1 and we speak of a *crossover region* there. The combined

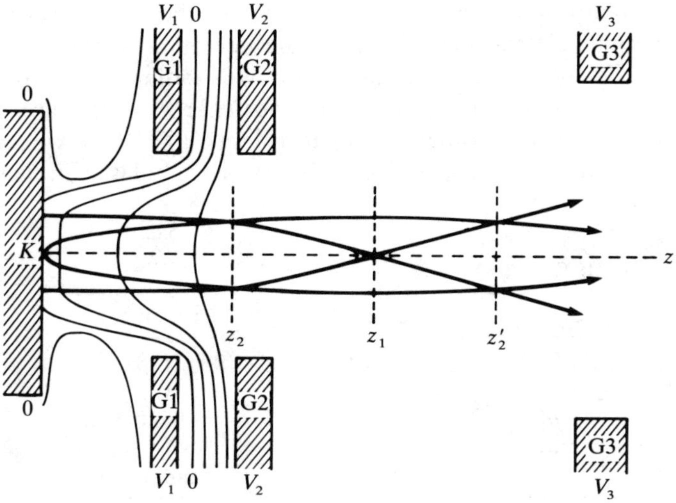

Fig. 11-3. An actual beamforming region with some equipotential planes and several typical electron trajectories.

effects of lens aberrations, Coulomb (space charge) repulsion of electrons by other electrons, and thermal distribution of emitted electrons smear the crossover out to a region rather than to a well-defined point. Application of the Sturm invariant (Eq. (6.2)) enables us to estimate simply the magnitude of the radius ρ_{cr} of the spot at the crossover:

$$\sqrt{\Phi(z)}[\rho_1'(z)\rho_2(z) - \rho_1(z)\rho_2'(z)] = \text{constant}\,. \tag{11.3}$$

The two trajectories implied by $\rho_1(z)$ and $\rho_2(z)$ are chosen as shown in Fig. 11-4. A principal electron with energy ~0.1 eV emitted at the rim of

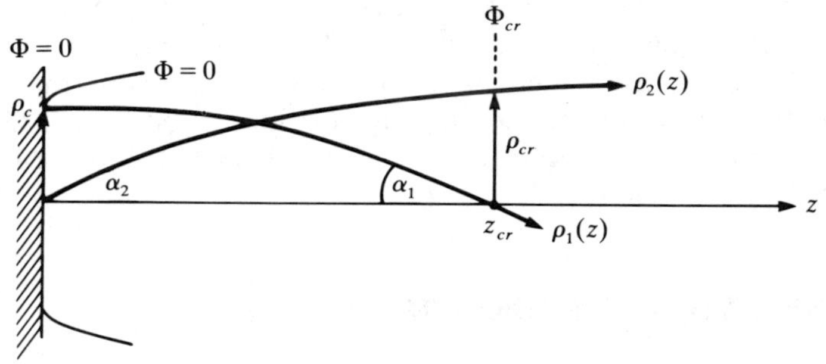

Fig. 11-4. A principal and a centrally emitted electron trajectory yielding an idealized image at the crossover in the BFR.

the cathode-emitting area (i.e., at $\rho = \rho_c$) traces the trajectory $\rho_1(z)$, and another thermal eletron emitted from the center at angle $\alpha_2 = \pi/4$ rad traces $\rho_2(z)$. These two produce an image with radius ρ_{cr} at crossover in the absence of aberrations and when space-charge effects are ignored. That will suffice for an estimate of ρ_{cr} at distance z_{cr}. When we apply Eq. (11.3) to $z = 0$ and $z = z_{cr}$, and use the values tabulated in Table 11-1 (see also Fig. 11-4),

TABLE 11-1

Energy	Cathode	Crossover
$e\Phi(z)$	$k_B T$	$e\Phi_{cr} + k_B T$
ρ_1	ρ_0	0
ρ_1'	0	$-\tan \alpha_1$
ρ_2	0	ρ_{cr}
ρ_2'	$\tan \alpha_2$	$\tan \tilde{\alpha}_2$

with the fact that $\tan \alpha_2 = 1$, we obtain

$$\rho_{cr} \sim \sqrt{\frac{k_B T}{e\Phi_{cr} + k_B T}} \frac{\rho_c}{\tan \alpha_1} = \sqrt{\frac{k_B T}{e\Phi_{cr} + k_B T}} f_{\mathrm{BFR}}, \tag{11.4}$$

where $\tan \alpha_1 = \rho_c/f_{\mathrm{BFR}}$ invokes the focal distance f_{BFR} of the lens formed by the cathode and the G1 and G2 grids. It is obviously advantageous to have Φ_{cr} as large as possible. However, it should be borne in mind that space charge has been ignored as well as aberrations. Space charge is a strong function of current strength; in fact we have seen (in Chapter 10) that the broadening due to space charge is governed by the ratio $I/\Phi_{cr}^{3/2}$ so that this effect enhances the importance of large Φ_{cr}, but the effect still depends upon the current strength.

Partly for these reasons (but also because a screen spot needs to be produced), the main lens places a G3 grid at a much higher voltage V_3 onto the BFR. Because the crossover region extends beyond the G2 grid it follows that the focal distance f_{BFR} is also influenced by the G3 lens. For that reason, the initial part of the G3 lens is included in the BFR, as shown in Figs. 11-1 and 11-3.

11.3 THE MAIN-LENS REGION (MLR)

The BFR produces a beam that enters the G3 lens as a divergent bundle of trajectories. It will be necessary to pass the beam through one more lens, possibly compound, to obtain a small focussed spot on an image plane (screen). The simplest type of lens consists of two coaxial consecutive cylinders: a G3 cylinder at V_3 volts, and a G4 cylinder at a considerably higher voltage, V_4. The electrons will have been accelerated to an energy eV_4, as shown in Fig. 11-5. The electrons then are no longer accelerated and they drift at a velocity proportional to $\sqrt{eV_4}$ in straight-line trajectories. If the G3 cylinder is sufficiently long, then it too will contain a field-free region in which electrons have a constant velocity vector. Electrons entering the G3 at different radial distances and angles will focus at different z-axis locations due to the combined effects of the initial thermal spread, the lens aberrations in the BFR and the MLR, and space charge (principally in the drift region as we have yet to see). If we ignore the aberrations (for rays not far from the axis) and space charge (for low current values), then the image smear on the screen is due only to thermal variations in electron velocities which we will now analyze separately.

If we consider the region common to the G3 and the G4 cylinders in which the lensing action takes place, then we note that electrons travel in straight lines immediately preceding and following that region. If we extrapolate the entering straight-line pieces of trajectory back towards the cathode then they will intersect the axis at a range of points that usually extends well into the cathode and beyond. There will be some location where this backwards-extended beam is narrowest; we call this the *virtual cathode* which is located at a distance L_0 from the object principal plane P_3. Likewise, the screen is at a distance L from the image principal plane P_4; see Fig. 11-6 for a sketch of the paraxial optics of this situation.

The argument leading to Eq. (11.4) can now be repeated but with the screen serving as the image plane. It yields

$$\rho_{th} \sim \sqrt{\frac{k_B T}{eV_4 + k_B T}} \frac{\rho_c}{\tan \alpha_s}, \tag{11.5}$$

under the assumption that we are imaging thermal electrons with an energy spread $\sim k_B T$. Recognizing that $k_B T \ll eV_4$ and $\tan \alpha_s = \rho_B / L$ (see Fig. 11-5), we obtain

$$\rho_{th} \sim \sqrt{\frac{k_B T}{eV_4}} \frac{L}{\rho_B} \rho_c . \tag{11.6}$$

These last two equations would have "=" signs connecting left and right sides instead of "~" signs if they had been derived from Eq. (11.3) for a

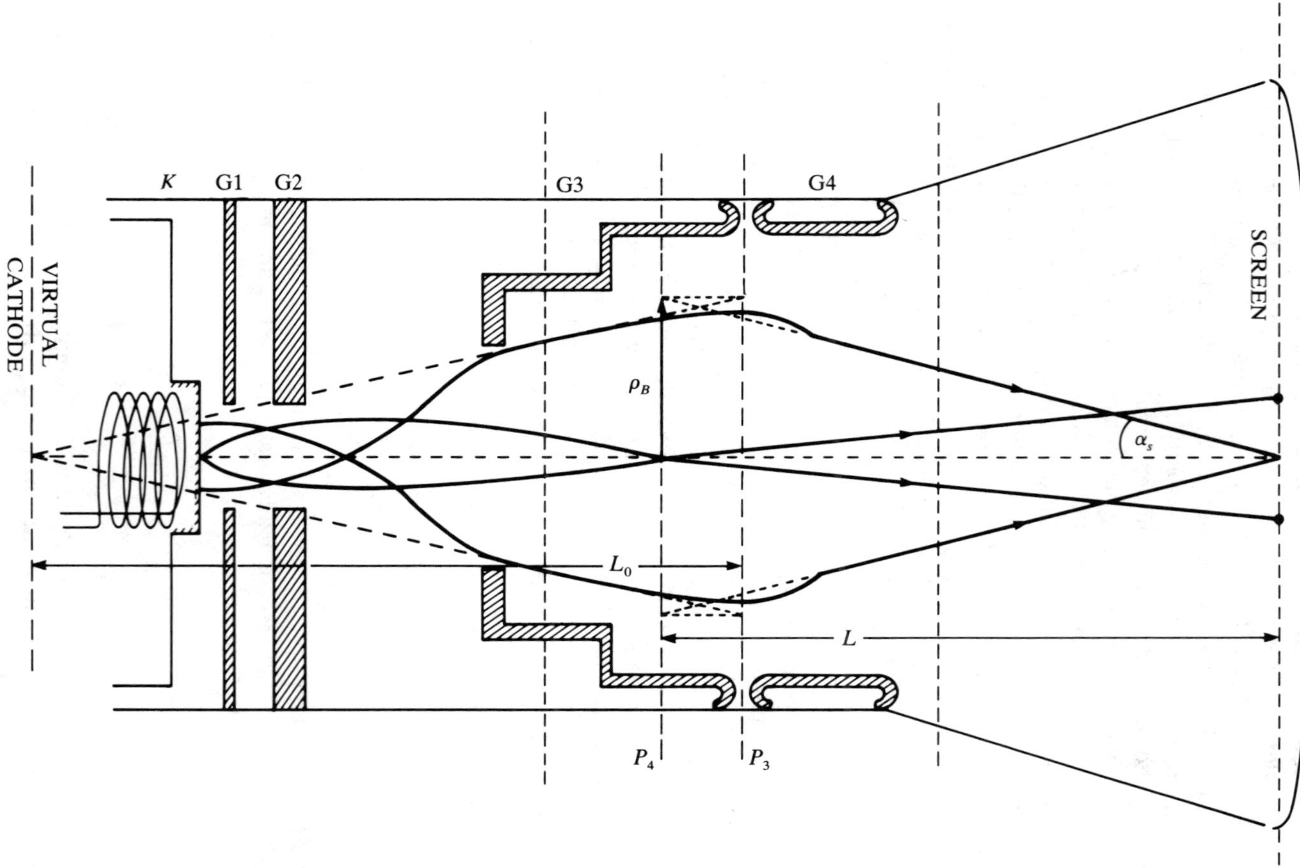

Fig. 11-5. Repeat of Fig. 11-1 with some typical trajectories followed from the cathode to the screen.

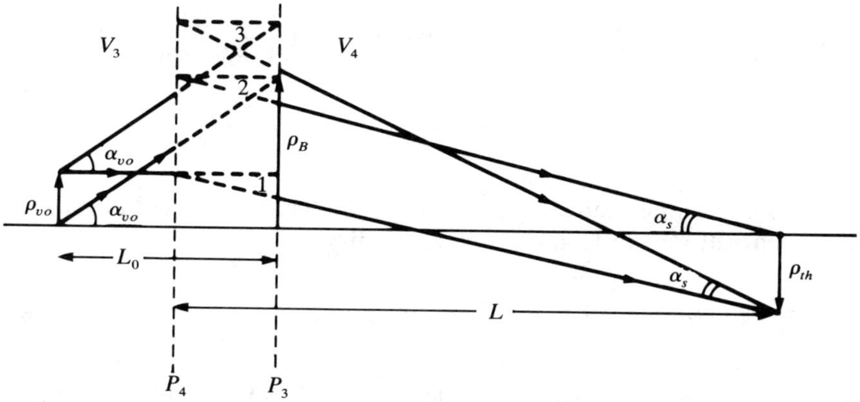

Fig. 11-6. A paraxial-optics sketch of rays through the main lens, extended backwards to a virtual crossover point, and forwards to a thermal spot on the screen.

principal electron emitted with energy $k_B T$ and for an axial electron, also at $k_B T$, emitted at angle $\tilde{\alpha}_2 = 90°$ but landing at radius ρ_{th} at the screen. We need to qualify this as an approximate argument because we are ignoring the details of the thermal distribution. At any rate, there will be some effective radius ρ_{th} for which Eq. (11.6) is exact, provided the beam radius ρ_B is suitably defined. The sketch in Fig. 11-6 shows that ρ_B is a measure of the maximum radius of the beam in the lens.

The present purpose of introducing Eq. (11.6) is to provide us with a means of estimating the "throwlength" L. To do so, we note that the radius of the emitting area of a cathode at maximum current is $\rho_0 \sim 4$ mm, whereas the ratio $\sqrt{k_B T/eV_4} \sim 0.0018$ for a 30 kV G4 cylinder lens. Because the actual physical radius of an electron gun in a TV kinescope is of the order of 1–2 cm, we estimate that $\rho_B \leqslant 7.5$ mm. Because it is desired to have ρ_{th} less than 1 mm, these rough estimates indicate that $L \sim 1$ m. This is unacceptably large of course, and various improvements in design help bring this length down to less than 40 cm in practice, but the main conclusion of this type of argument is that the main lens cannot bring a BFR-generated beam of electrons to a reasonable focus within a distance of 30–50 cm beyond the ML unless quite different designs are employed. This gives rise to the *drift region* (to be discussed below).

Moss (1968) discusses and extends calculations by Pierce (1954) of charge density at the screen. We cite his result for current density at the screen without derivation:

$$J_S = \frac{1}{M^2} J_0\left[1-(1-M^2\sin^2\alpha_s)\exp\left\{-\frac{eV_4}{k_BT}\left(\frac{M^2\sin^2\alpha_s}{1-M^2\sin^2\alpha_s}\right)\right\}\right], \tag{11.7}$$
$$M \equiv \frac{\rho_{th}}{\rho_0}.$$

In the case that the inequality $M \sin \alpha_s \ll 1$ holds, the current density at the screen will reduce to the Langmuir limit,

$$J_s = J_0\left[\frac{eV_4}{k_BT} - 1\right]\sin^2\alpha_s, \tag{11.8}$$

where $\sin \alpha_s \approx \rho_B/L$ is always much less than unity. The condition for Eq. (11.8) is usually met in practice because magnification M is usually not sufficiently large to compensate for the small size of $\sin \alpha_s$, and thus this value of screen-current density serves as a measure of *spot brightness*. Brightness is enhanced by choice of large V_4 and large α_s (i.e., large beam radius ρ_B). Both V_4 and ρ_β are controlled by main-lens design.

An inhibiting factor in main-lens design is the unavoidable presence of *aberrations* due to the conflicting need, indicated in Eq. (11.6), of making ρ_B as large as possible. In Section 8.1.2 we showed that the aberration broadening at the image plane is given by

$$\rho_{ab} = C_{ia}\alpha_{v0}^3, \tag{11.9}$$

referring to Fig. 11-6 for definition of angle α_{v0}, under the assumption that the angle is sufficiently small so that $\tan \alpha_{v0} \approx \alpha_{v0}$. Here, C_{ia} is the aberration coefficient for imaging, and it requires knowledge of a reference trajectory that originates at the virtual cathode at an angle of 45°, see Eq. (8.23). A glance at the geometry shows that

$$\rho_B \approx L_0\alpha_{v0} \approx L\alpha_s, \tag{11.10}$$

so that we obtain

$$\rho_{ab} = C_{ab}\rho_B^3, \tag{11.11}$$
$$C_{ab} = -M\frac{1}{L_0^3}\int_{-z_{vc}}^{L} d\zeta\sqrt{\frac{\Phi(\zeta)}{V_3}}\,\rho_2(\zeta)\mathcal{R}_3(\rho_2, \rho_2'; \Phi_0),$$

in the understanding that $\rho_2(\zeta)$ is the above-mentioned 45° trajectory. We observe that aberration broadening is proportional to ρ_B^3, whereas thermal broadening is proportional to ρ_B^{-1}.

Main-lens design is therefore dominated by the desire to reduce the aberration coefficient C_{ab} to a minimum. We can always reduce $C_{ab}\rho_B^3$ by reducing C_{ab}, i.e., by designing a lens with fixed throwlength L and magnification M but with reduced aberration coefficient C_{ab}. That would then allow us to increase ρ_B so as to reduce ρ_{th} (see Eq. (11.6)). However, the need to restrict the total diameter of an electron gun imposes a manufacturing constraint upon the magnitude of ρ_B.

These arguments need to be extended to include broadening effects due to space charge. That effect is largest in the *drift region* (DR), which is discussed separately in the next section.

11.4 THE DRIFT REGION (DR)

We have just observed (in Section 11.3) that a rather large throwlength $L \leqslant 0.5$ m is needed to allow a main lens to focus a beam produced in the BFR without excessively high V_4 voltage. It should be stressed here that the entire drift region is at one potential value, V_4, which is known as the *ultor* voltage in a CRT. That does not imply that the MLR consists of only two cylinders held at the potentials V_3 and V_4. The main lens may be quite compound with many intermediate grids or cylinders, but the convention we adapt here is to label the initial part of the lens G3 (with voltage V_3) and the final part G4 (with voltage V_4).

With reference to Fig. 11-7, consider an electron that leaves the MLR

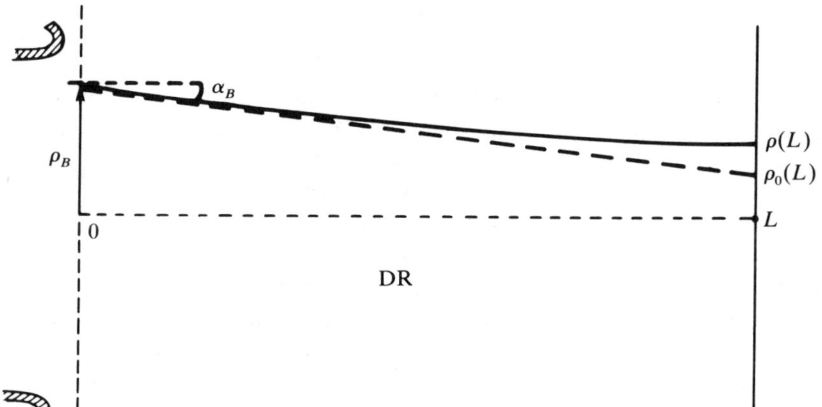

Fig. 11-7. Illustration of the effect of space charge upon a trajectory in the drift region.

at radial distance ρ_B and at an angle $-\alpha_B = \rho'_B$ with the horizontal axis. We assume that $\alpha_B \ll 1$ so that $\tan \alpha_B \approx \alpha_B$. We showed in Eq. (10.13) that the trajectory is then governed by the equation

$$\frac{d^2\rho}{dz^2} = \frac{K}{2\rho}, \tag{11.12}$$

with $K = \beta I/V_4^{3/2}$, and I is the current inside $\rho = \rho_B$ under the assumption of laminar annuli of current. If $K = 0$ then this equation yields a straight-line trajectory with

$$\rho_0(L) = \rho_B + L\rho'_B. \tag{11.13}$$

We shall work out an approximate solution for the space-charge case $K \neq 0$ under the assumption that $|\rho(L) - \rho_0(L)| \ll |\rho_0(L)|$, i.e., that the space-charge does not modify the location of $\rho_0(L)$ greatly. That assumption is easily seen to yield the following approximate solution of Eq. (11.12):

$$\rho(L) = \rho_B + \frac{K}{4\rho_B} L^2, \tag{11.14}$$

and consequently the radius of the spot is broadened due to space-charge repulsion by an amount,

$$\rho_{sc} = \frac{KL^2}{4\rho_B}. \tag{11.15}$$

There are a number of reasons why this expression is at best an approximation. First an empirical reason: an evaluation for a typical "highlight" current of 3.5 mA in a 30 kV drift region ($K \sim 2 \times 10^{-5}$), and for $L \approx 34$ cm and $\rho_B \sim 1$ mm, yields a space-charge radius $\rho_{sc} \sim 0.6$ mm, which is not small enough for validity of the approximate solution (11.14). In the second place, electron trajectories do cross each other, as can be seen in the sketch of Fig. 11-8 for a number of principal electrons emitted from the cathode. In spite of these limitations and violations of the strict conditions of validity, Eq. (11.15) does provide an order-of-magnitude estimate of space-charge broadening. The proportionality of ρ_{sc} to the factor L^2/ρ_B clarifies why space-charge broadening occurs most strongly in the drift region. Only in the crossover region of the BFR is there possibly a situation in which the large magnitude of $1/\rho$ offsets the small value of z^2 in the factor z^2/ρ that would hold there.

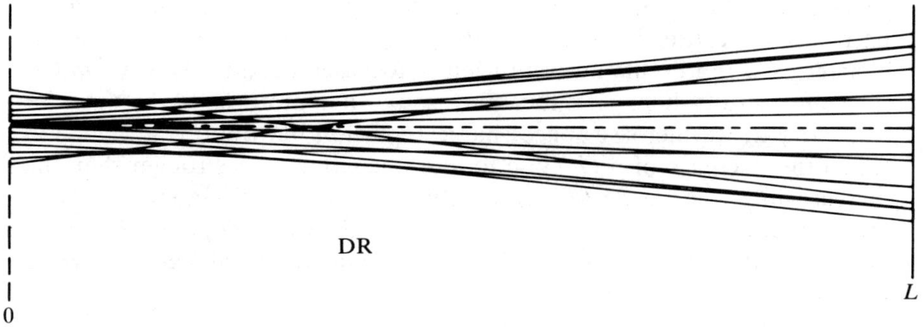

Fig. 11-8. Electron trajectories between the main lens and the screen in a TV kinescope. Note crossings of trajectories.

11.5 FOCUSSING OF THE BEAM INTO A SCREEN SPOT

11.5.1 Phase-Space Representation

The strategy for achieving best focus is complicated in view of the fact that there are three effects that oppose that goal: *thermal broadening* (Eq. (11.6)); *aberration broadening* (Eq. (11.11)); and *space-charge broadening* (Eq. (11.15)). Nevertheless, there is a strategy, and this section will prepare us for dealing with it. Let us first assume that the beam diameter in the MLR is ρ_B and that it is fixed in magnitude. A trajectory that represents the edge of a beam in the DR would be represented by one of the three depicted in Fig. 11-9. Trajectory 1 is of an electron aimed at a focus on the axis at some distance $z \leqslant L$, and

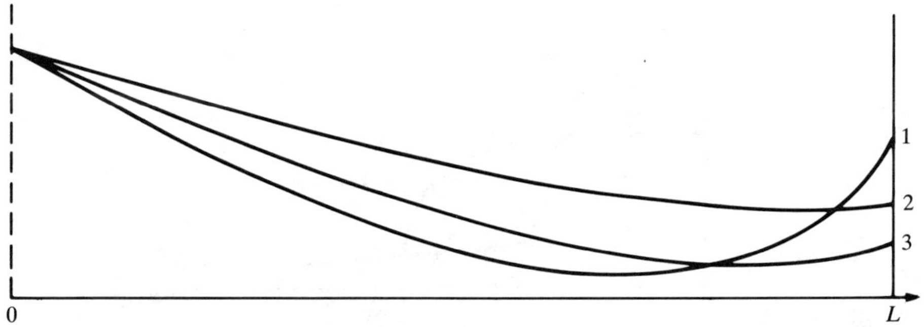

Fig. 11-9. Determination of a best-focus strategy. Note the effects of space charge on overfocussing.

space-charge broadening due to concentration of charge carriers inside a narrow waist causes it to fan out excessively; *overfocussing* is the result. Trajectory 2 represents an insufficiently focussed edge trajectory; *underfocussing* results here. By focussing somewhat more strongly than for trajectory 2, we increase space-charge repulsion but there is a regime in which the decrease of $\rho_0(L)$ due to more negative ρ_B' is stronger than the increase in ρ_{sc} and a *best-focus* situation can arise (trajectory 3). These three situations are also shown in Fig. 11-10 which depicts *slit-current profiles* at the screen ($z = L$). The slit-current density is defined by the integral,

$$J_s(\rho) \equiv \int_{-\infty}^{\infty} dx\, J(\rho, L)\,, \qquad \rho \equiv \sqrt{x^2 + y^2}\,. \tag{11.16}$$

This definition corresponds to passing a uniform slit detector over a more or less circular spot. The slit-current density has the property of being maximal in general at $\rho = 0$, unlike $J(\rho, L)$ which is often maximal at some nonzero value of ρ. However, the main features illustrated in Fig. 11-10 for the slit-current profiles would be quite similar for the ordinary current density profiles.

Figures 11-11 show sketches of the actual phase-space evolution (see Chapter 9) of a typical kinescope beam (Washino et al. 1979). In part (a) we see a phase space occupied by electrons with energy less than 0.1 eV emitted from all locations that can radiate electrons out of the cathode ($\rho < \rho_0$). This is a rectangular phase-space area but the occupation by electrons in it is not uniform; the density is given by $F_0(\vec{\rho}, \vec{p}_T)$. It is not

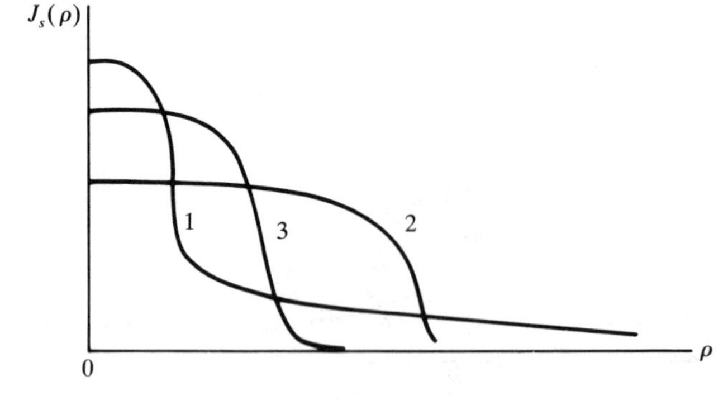

Fig. 11-10. The slit-current density profiles for the three situations depicted in Fig. 11-9.

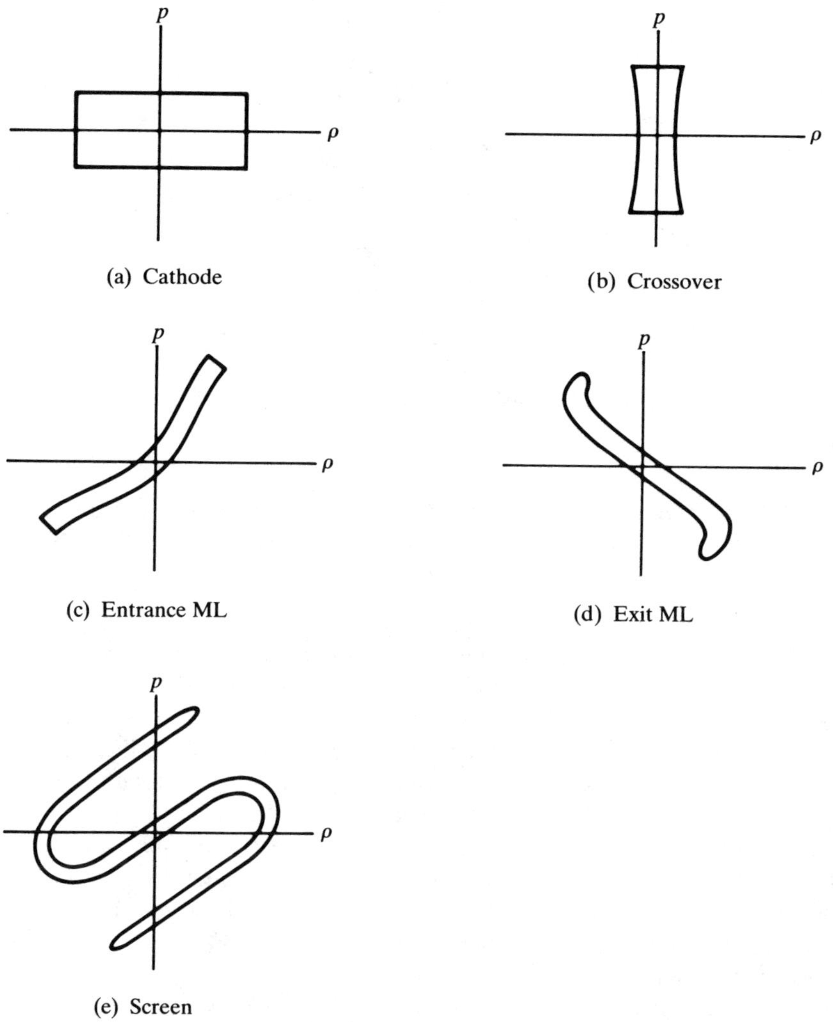

Fig. 11-11. The ρ-p phase space area occupied by most of the electron beam at various locations between cathode and screen. Note the increased effect of aberrations and space charge at further distances.

necessary to take the phase-space density into account here. As the beam progresses towards the crossover region the principal electrons will indeed tend to focus (if aberrations and space-charge repulsion are ignored), but electrons emitted at sizable angles with the normal will tend to be more or less parallel to the axis. As a result, the outline of the

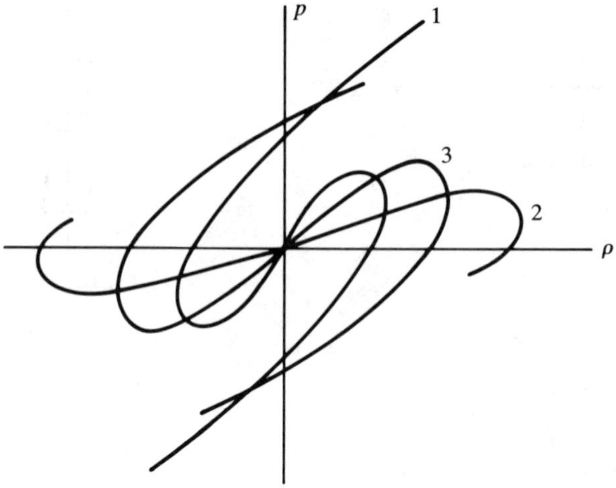

Fig. 11-12. The phase-space picture at the screen for a row of principal electrons emitted from the cathode each line is for one of the situations of Fig. 11-9.

phase-space area (which itself should be conserved!) will rotate by 90° as shown in Fig. 11-11b. The further development shown in parts (c)–(e) of Figs. 11-11 can be made plausible by following trajectories such as those sketched for the DR in Fig. 11-8. Aberrations cause the S-like curling of the phase-space areas that increases from (c) to (e) in these figures. Note that the phase-space area shown in part (e) is for a best-focus situation. The situations of over- and underfocussing are illustrated in Figs. 11-12, which demonstrate what happens to an initial straight line of principal electrons emitted from the cathode when they reach the screen; the trajectory numbers refer to the situations of Figs. 11-9 and 11-10. Each initially straight line is stretched and curled by aberrations. The slope close to the axis is determined by the focussing strength (parameter S of Eq. (6.18)), and even fairly small changes in S rotate the figure.

11.5.2 Phenomenological Best-Focus Theory

A phenomenological theory of lens design can be based upon the three broadening effects given in Eqs. (11.6), (11.11), and (11.15). We can summarize the theory succinctly with a rather simple equation by stating that the radius of the screen spot is given approximately by

$$\rho_s = C_1 \rho_B^{-1} + C_3 \rho_B^3 \,. \tag{11.17}$$

The coefficient C_1 (dimension m^2) represents some combination of thermal and space-charge broadening. Many authors advocate a Pythagorean sum,

$$C_1^2 = a_T \frac{k_B T}{eV_4} L^2 \rho_c^2 + \alpha_{sc} \frac{K^2 L^4}{16} , \qquad (11.18)$$

where coefficients a_T and a_{sc} are constants of order unity which are chosen in some phenomenological way (Moss 1968, Harte 1973, Wilson 1975). These constants depend upon how one defines ρ_B and ρ_c; each radius can be a 5% measure (i.e., 5% of the maximum value), or a $1/e$ measure, or a root-mean-square, etc. Furthermore, the reasoning given for a Pythagorean sum is not convincing and a linear sum, for example, may be just as useful. The real point is that $C_1 \rho_B^{-1}$ must reduce either to ρ_{th} or to ρ_{sc} when either of the two terms in Eq. (11.18) is dominant. The same caveat should be applied to Eq. (11.17) where a linear sum has been used arbitrarily; it is merely convenient for phenomenological analysis and any other properly specified combination which reduces to the correct term when it is the dominant effect will do just as well. We are not overly concerned about the precise addition theorem, which is bound to be complicated and different for different tubes, because we will only draw a number of general design conclusions that in all likelihood do not depend on these details. A reminder: we have already learned that the dimension of C_3 is m^{-2}.

Figure 11-13 shows a plot of $\rho_s(\rho_B)$ as a function of ρ_B for fixed values of coefficients C_1 and C_3. There is a minimum value for the spot radius, $\bar{\rho}_s$, and it occurs when the beam radius is $\bar{\rho}_B$ in the MLR. It follows immediately from Eq. (11.17) that

$$\bar{\rho}_B = C_1^{1/4}(3C_3)^{-1/4} , \qquad \bar{\rho}_s = \tfrac{4}{3} C_1^{3/4}(3C_3)^{1/4} . \qquad (11.19)$$

These values also yield $\bar{\rho}_B \bar{\rho}_s = \frac{4}{3} C_1$ so that the optimal spot radius is 4/3 times what it would have been for an aberration-free lens with the same beam diameter $\bar{\rho}_B$. Improvements can be made only by modifying the coefficients C_3 and C_1. In fact, modification of C_1 is probably not possible without drastic improvements in the BFR design or introduction of a significantly improved cathode emitter. Modification of C_3 implies a redesign of the main lens, as we shall see shortly. If, however, modification of these two parameters is not feasible for some reason (time, cost, etc.) then it is clearly preferable to *underoptimize* ρ_s (i.e., to design at a beam radius ρ_B than is somewhat *smaller* than the optimum $\bar{\rho}_B$. The reason is obvious upon inspection of Eq. (11.17): whatever the precise

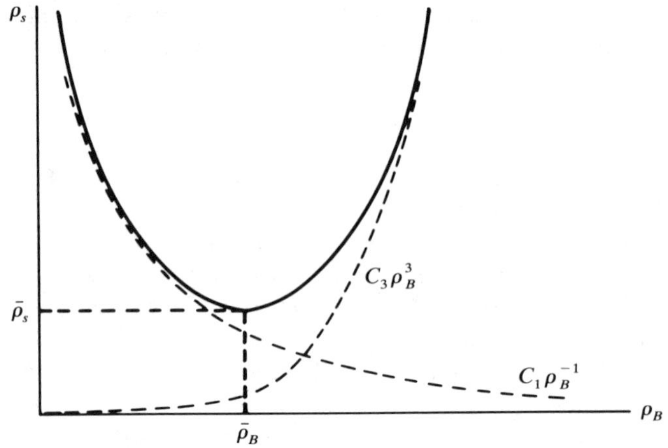

Fig. 11-13. Spot radius ρ_S versus beam radius ρ_B with optimized curve as sum of two dashed curves due to aberrations ($\propto \rho_B^3$) and to space charge + thermal effects ($\propto \rho_B^{-1}$).

addition theorem may be, any small change in ρ_B at underoptimization manifests itself as a change in ρ_s proportional to $1/\rho_B$ rather than to ρ_B^3.

Proper use of the optimization based upon these simple phenomenological formulas usually requires a computer simulation of the entire electron gun and the beam through it. Minor adjustments in MLR voltages usually require a new simulation which will yield a new spot radius ρ_s at a new beam radius ρ_B. Since complete beam simulation is somewhat costly in computer time, it is highly desirable to restrict the number of such simulations to a minimum. If one can manage to do several simulations (three suffice) that lie close to the optimum, then a regression analysis based on Eq. (11.17) should yield C_1 and C_3, and, with Eq. (11.19), the optimal values $\bar{\rho}_B$ and $\bar{\rho}_s$. Further optimization requires a decrease in the coefficients C_3 and/or C_1, as stated above. If successful, that would produce a new curve of ρ_s vs ρ_B on Fig. 11-13 that lies below the older one, and thus allows us to optimize at a lower $\bar{\rho}_s$. The concomitant new $\bar{\rho}_B$ will be *larger* if coefficient C_3 has been made smaller, and *smaller* if coefficient C_1 has been reduced (see Eqs. (11.19)).

11.5.3 Optimization of Aberrations in the MLR

The first and most obvious way to improve the optimal spot radius $\bar{\rho}_s$ is to redesign the main lens so that its focal strength (parameter S) remains essentially unchanged, but so that the aberrations are decreased. That implies that we attempt to reduce parameter C_3 in Eq. (11.17). A reasonable estimate for C_3 is the *aberration coefficient* C_{ia} given in Eq. (8.23) for image-plane aberrations. If we replace the small angle θ_0 by ρ_B/L_0 (see Fig. 11-5 to infer what angle θ_0 is) we obtain

$$C_3 \approx -\frac{M}{L_0^3}\int_0^\infty d\zeta \sqrt{\frac{\Phi_0(\zeta)}{V_3}}\; \rho_2(\zeta)\mathscr{R}_3(\rho_2, \rho_2'; \Phi_0)\,, \qquad (11.20)$$

where $\rho_2(\zeta)$ is the (hypothetical) 45° trajectory emitted from the axis at the virtual-cathode location (where the potential can be thought to be V_3), and M is the magnification defined as $M = (L - f)/f = f_0/(L_0 - f_0)$. The factor $\mathscr{R}_3(\rho_2, \rho_2'; \Phi_0)$ was derived in Eq. (8.5) and it contains some terms proportional to $\Phi_0^{(3)}$ and $\Phi_0'\Phi_0''$ which are modified after integration by parts. Following Grivet (1972), we then obtain

$$C_3 \approx -\frac{M}{L_0^3}\int_0^\infty d\zeta \sqrt{\frac{\Phi_0(\zeta)}{V_3}}\; [\rho_2(\zeta)]^4\mathscr{R}_4(\rho_2, \rho_2'; \Phi_0)\,, \qquad (11.21a)$$

$$\mathscr{R}_4(\rho_2; \rho_2', \Phi_0) = \frac{1}{32}\left[\frac{\Phi_0^{(4)}}{\Phi_0} - \left(\frac{\Phi_0''}{\Phi_0}\right)^2\right] - \frac{1}{4}\frac{\Phi_0''}{\Phi_0}\left(\frac{\rho'}{\rho}\right)^2 - \frac{1}{2}\left(\frac{\rho'}{\rho}\right)^4 .$$

An alternative form due to Glaser (1952) for the $\mathscr{R}_4$ factor is

$$\mathscr{R}_4(\rho_2; \rho_2', \Phi_0) = \frac{1}{64}\left[3T^4 - \frac{9}{2}\,T^2T' + 5(T')^2\right] + \frac{1}{16}\,TT'\,\frac{\rho'}{\rho}\,, \qquad (11.21b)$$

where $T(\zeta) \equiv \Phi_0'(\zeta)/\Phi_0(\zeta)$. This form has the advantage that it shows that $\mathscr{R}_4 = 0$ whenever the prime derivatives of $\Phi_0(\zeta)$ vanish, i.e., in drift regions because it is zero when both T and T' are zero. In other words, it shows that $\mathscr{R}_4(\rho_2, \rho_2'; \Phi_0)$ is nonzero only in lensing regions.

The transition from Eq. (11.20) to Eqs. (11.21) requires integration by parts, and the latter equations are therefore valid only if the initial and final z values lie in field-free regions. That is the case in the application to main lens and drift region under consideration as can be seen by perusing Fig. 11-6 again.

Several useful approximations can be made. First, the $(\rho')^2$ and the $(\rho')^4$ terms in Eq. (11.21a) are effectively very small because ρ'/ρ is a very small quantity in the lensing region. Integration by parts of Eq. (11.21b) then shows that $\mathscr{R}_4$ is effectively given by

$$\mathscr{R}_4(\rho_2; \rho_2', \Phi_0) = -[\tfrac{3}{8}T^4 + \tfrac{3}{2}(T')^2]\,, \qquad (11.21c)$$

which indicates that C_3 must be positive-definite and is therefore minimized by making it zero. Scherzer (1936) has given a more general proof that C_3 is positive-definite. Hence we obtain

$$C_3 \approx \frac{M}{32L_0^3} \int_0^\infty d\zeta \sqrt{\frac{\Phi_0(\zeta)}{V_3}} \left[\left(\frac{\Phi_0''}{\Phi_0} \right)^2 - \frac{\Phi_0^{(4)}}{\Phi_0} \right] [\rho_2(\zeta)]^4 \qquad (11.22)$$

It is useful to replace $1/\sqrt{V_3}$ by $f/(f_0\sqrt{V_4})$, using Eq. (6.11), because we intend to modify the main lens by adjusting V_3 but maintaining the ultor voltage V_4. We may also approximate $\rho_2(\zeta)$ by L_0 to obtain

$$C_3 \approx \frac{MfL_0}{32f_0} \int_0^\infty d\zeta \sqrt{\frac{\Phi_0(\zeta)}{V_4}} \left[\left(\frac{\Phi_0''}{\Phi_0} \right)^2 - \frac{\Phi_0^{(4)}}{\Phi_0} \right]. \qquad (11.23)$$

Optimization (i.e., attainment of a minimum) of C_3 is then realized by an effort to make the factor in square brackets in this equation as small as possible. In fact, the factor can be made to be precisely zero by choosing $\Phi_0(z)$ to be one of the following three axial-potential profiles:

$$\begin{aligned} \Phi_0(z) &= a_1 + b_1 z , \\ \Phi_0(z) &= a_2 \exp\left(\frac{z}{\lambda}\right) + b_2 \exp\left(-\frac{z}{\lambda}\right), \\ \Phi_0(z) &= a_3 \sin\left(\frac{z}{\lambda}\right) + b_3 \cos\left(\frac{z}{\lambda}\right), \end{aligned} \qquad (11.24)$$

where coefficients a_i, b_i, and λ are fixed parameters. The choice of an axial-potential profile is dictated partly by the boundary conditions that $\Phi_0(z) \rightarrow V_3$ as $z \rightarrow 0$, and $\Phi_0 \rightarrow V_4$ as $z \rightarrow \infty$. None of the three profiles of Eq. (11.24) satisfy these boundary conditions by themselves, but a realistic profile of Φ_0 might consist of various pieces chosen from each of these three and joined together as smoothly as possible. It is obvious that a "smooth" joint (all derivatives of Φ_0 exist and are finite) is not always possible but this strategy should be able to yield low values of C_3. Figure 11-14 shows the axial profile of a typical "bi-potential" main lens consisting of two co-axial cylinders with equal radius and with a small gap between them such that $V_4 > V_3$. The lower figure gives a qualitative sketch of the function $\Phi_0''(z)/\Phi_0(z)$, which is the dominant part of the integrand in Eq. (11.23) when squared. The contribution to this ratio clearly comes from the locations where the zero-slope rising part joins on to the asymptotic values of $\Phi_0(z)$. The other parts contribute negligibly because they closely resemble parts of the profiles of Eq. (11.24). Other

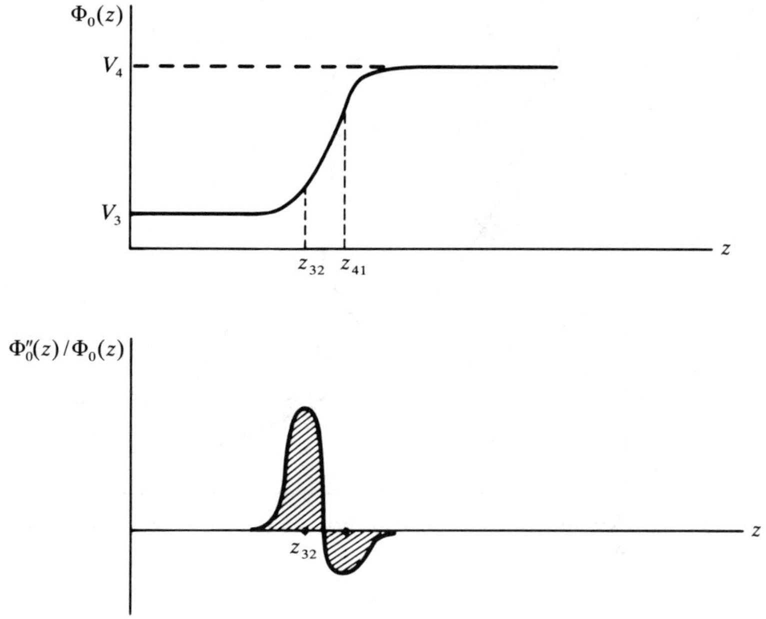

Fig. 11-14. Axial profile of $\Phi_0(z)$ and of the ratio $\Phi_0''(z)/\Phi_0(z)$ for a bipotential lens.

types of main lenses include various kinds of "tri-potentials". The best known is the *unipotential* lens (formerly known as the "einzel" lens) which consists of three equi-radius co-axial cylinders (separated by small air gaps) with potentials V_3, V_m, and V_4 with $V_4 = V_3$ and with $V_m < V_3$. Another variant has $V_4 > V_3$, and $V_m \sim \sqrt{V_3 V_4}$ has been chosen in an attempt to model the rising part by an exponential profile.

The procedure by which we attempt to optimize C_3 is illustrated in Fig. 11-15. The long curve is a plot of ρ_s vs ρ_B subject to the condition that $\rho_s \rho_B = C_1$, which is what would be the case if an aberration-free ML were to be placed beyond the BFR. That is a kind of "ideal-lens" case with which all real lenses are to be compared. The short curves labeled 1, 2, and 3 are graphical plots of Eq. (11.17) for three main lenses with aberration coefficients that decrease from #1 to #3. We note that each lower-aberration lens yields an optimal $\bar{\rho}_s$ at a $\bar{\rho}_B$ that has a larger value than for a higher-aberration lens. We also note that the locus of the minimum values $\bar{\rho}_s(\rho_B)$ as a function of ρ_B lies on the curve $\bar{\rho}_s = \frac{4}{3} C_1/\rho_B$. A significant decrease in C_3 by means of main-lens redesign allows the possibility of underoptimization of ρ_B without exceeding the $\bar{\rho}_s$ of a previous more highly aberrating ML design.

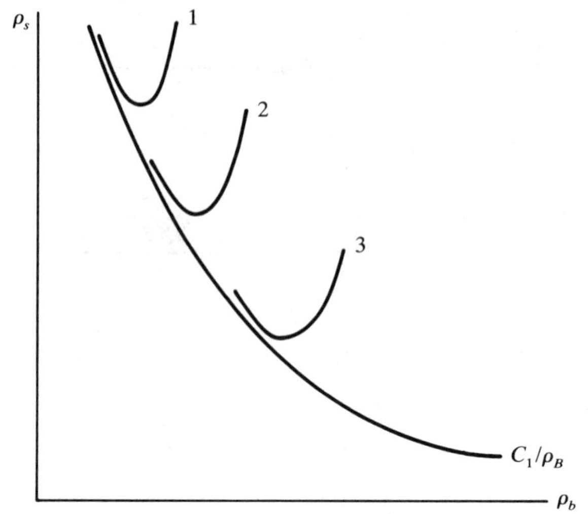

Fig. 11-15. Curves that illustrate optimization strategy or reducing coefficient C_3 for an electron gun.

The product $\rho_B \rho_s$ is sometimes regarded as a *figure of merit*. We note from Eq. (11.17) that the product $\rho_B \rho_s = C_1 + C_3 \rho_B^4$, hence that C_1 is a minimum value of this product. For a given lens with certain C_3, we observe that $\bar{\rho}_B \bar{\rho}_s = \frac{4}{3} C_1$, i.e., that the figure of merit lies at 75% of the optimally obtainable value. If it is also noted that $\rho_B = L \tan \theta = L\rho'$, then the ideal figure of merit is $C_1 = L\rho_B \rho_B'$. Thus $C_1/L = \rho_B \rho_B'$ which is the phase-space area of an aligned rectangle representing a bundle of electrons in the main lens to zero order. The figure of merit (times throwlength L) therefore also represents a zero-order phase-space area!

11.5.4 Optimization of the C_1 Coefficient

Attempts to minimize aberrations in the MLR affect the optimal spot radius $\bar{\rho}_s$ in proportion to the factor $C_3^{1/4}$, according to Eq. (11.19). This is relatively inefficient in comparison to the effect of being able to improve upon coefficient C_1 because $\bar{\rho}_s \propto C_1^{3/4}$. The problem in improving C_1 can be understood from Eq. (11.18) which shows how two different physical effects contribute in some additive fashion to C_1; obviously Eq. (11.18) is oversimplified. Let us examine what is involved. The space-charge contribution is not easily modified because L and K are fixed for given ultor voltage and throwlength. However, the quantity ρ_c in Eq. (11.18) is not the actual radius of the emitting area of the cathode but some effective radius that is partially determined by space-charge and aberrational smearing of the first crossover. It seems obvious that redesign of the BFR to produce an entry beam into the MLR that appears to emanate from a smaller virtual crossover will also produce a smaller C_1 coefficient and hence will further optimize $\bar{\rho}_s$. Unfortunately, few guidelines seem to be available at present for a systematic optimization of C_1.

Yet another problem lies in the fact that optimization at one current does not imply an optimum at another current. We have seen that $K \propto I$ and we will see that $\rho_c \propto I^{\kappa}$ (where $\kappa \sim 0.25$, as will be discussed in some detail in Section 11.6.1). It is then obvious that the first term in Eq. (11.8) is proportional to $I^{0.5}$ and the second to I^2. The relative weighting of the two terms therefore changes. Moreover, ρ_B is a function of I and so is $\bar{\rho}_B$. In practice, optimization is usually done at a representative high current (3–3.5 mA) because the reduction of $\bar{\rho}_B$ that occurs as current I is decreased will tend to underoptimize ρ_s at the lower currents (which is preferable, as we have stated above, to overoptimization at higher current levels; see the discussion following Eq. (11.19)).

11.6 CATHODE EMISSION

This section is devoted to a discussion of the principles involved in launching a beam of electrons of desired current strength into an electron gun or other electron optics device. For convenience, we restrict ourselves to an axisymmetric situation. Consider therefore specifically the initial part of the BFR, which we sketch in Fig. 11-16 for one half of a

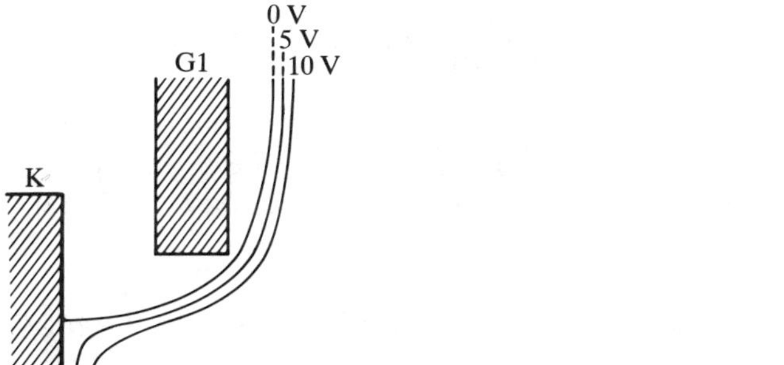

Fig. 11-16. Some equipotential planes in the beamforming region of an electron gun showing the formation of a diode-like region.

planar cut through the axis. Three equipotentials at $\Phi = 0$, 5, and 10 V are shown; the idea here is that 10 V is a relatively small increase in potential compared to the voltage $V_2 - V_1$. Close to the axis, the equipotentials are more or less planar. The deviation from a planar structure occurs at larger radii for values of Φ closer to the cathode voltage $V_K = 0$. If the emission is space-charge limited, then the potential for very small z and for $\rho < \rho_c$ (where ρ_c is the radius defined by the intersection of the $\Phi = 0$ equipotential with the cathode surface) is given by the theory of Sections 10.4 and 10.5,

$$\Phi(\rho, z) = \left(\frac{9\pi\beta z^2}{2}\right)^{2/3} [J(\rho)]^{2/3} , \tag{11.25}$$

where we remind the reader that $\beta = \sqrt{m/2e}/2\pi\epsilon_0 \approx 30.308\ \mathrm{V}^{3/2}/\mathrm{mA}$, and $J(\rho)$ is the space-charge limited current density (which is not initially a function of z). The reason that $J(\rho)$ becomes dependent upon z, and should therefore at a later stage be characterized as $J(\rho, z)$, is because the electrons are highly accelerated as they progress into the BFR. Their density decreases significantly and the beam is then no longer space-charge limited. However, immediately beyond the cathode, it is space-charge limited (except possibly at very low emission currents), and thus the current density has the dependence

$$J(\rho) \propto [\Phi(\rho, z)]^{3/2} , \tag{11.26}$$

and because $\Phi(\rho, z)$ is proportional to the Bessel function $J_0(2\rho/\rho_c)$, as demonstrated in Section 10.5, we find for $\rho < \rho_c$ that approximately

$$J(\rho) \approx J(0)\left(1 - \frac{\rho^2}{\rho_c^2}\right)^{3/2} . \tag{11.27}$$

Figure 10-6 clarifies that ρ_c is very close to the radius at which indeed $J(\rho) = 0$. Then we can obtain $J(0)$ from the integral for total current,

$$I = \int\int d^2\rho\ J(\rho) = 2\pi J(0) \int_0^\infty d\rho\ \rho\left(1 - \frac{\rho^2}{\rho_c^2}\right)^{3/2} = 0.4\pi\rho_c^2 J(0) , \tag{11.28}$$

so that the current density and potential in the space-charge limited region immediately beyond the cathode are given by

$$J(\rho) \approx \frac{5I}{2\pi\rho_c^2}\left(1-\frac{\rho^2}{\rho_c^2}\right)^{3/2}, \qquad \Phi(\rho, z) \approx \left[\frac{45\beta}{4}\frac{Iz^2}{\rho_c^2}\left(1-\frac{\rho^2}{\rho_c^2}\right)\right]^{2/3}. \tag{11.29}$$

We need to know how $\rho_c(I)$ depends upon the emitted current I in order to be able to solve equations of motion using Eq. (11.29) for a given emission current. We can observe this experimentally, or try to find this numerically. In the latter case, we can use weak-current (numerical) Laplace solutions for potential $\Phi(\rho, z)$ in the BFR near the cathode because even though the potential would be incorrect near the axis (where the current is space-charge limited and hence modifies the Laplace potential appreciably), it would be approximately correct in the vicinity of $\rho \sim \rho_c$ where the current density is low. The remaining problem, then, is to estimate emission current I for a given value of voltages V_1 on the G1 grid and V_2 on G2.

11.6.1 The Drive Characteristic

We need to understand how the voltages V_1 and V_2 in the BFR are chosen to obtain a desired emission current. This is done by establishing a *drive characteristic*, i.e., a plot of $I(V_1)$ vs V_1 which enables us to manipulate V_1 and thus to modulate the current I. We show the *cutoff* situation for $V_1 = -150$ V in Fig. 11-17. The term "cutoff" implies that all current emission is just shut off when the voltage on the G1 grid is decreased to the value $-V_{c0}$, and we have chosen $V_{c0} = 150$ V (a common choice in television electron guns) here. The values of V_2 and V_3 play a role here too because these voltages influence the potential profile closer to the cathode. Actually there is a minor problem here should we redesign the main lens and modify V_3. Even though this potential is shielded strongly from the BFR by the G2 grid with its V_2 voltage, it does have a minor influence, and a redesign of the main lens implies that V_2 and V_3 have to

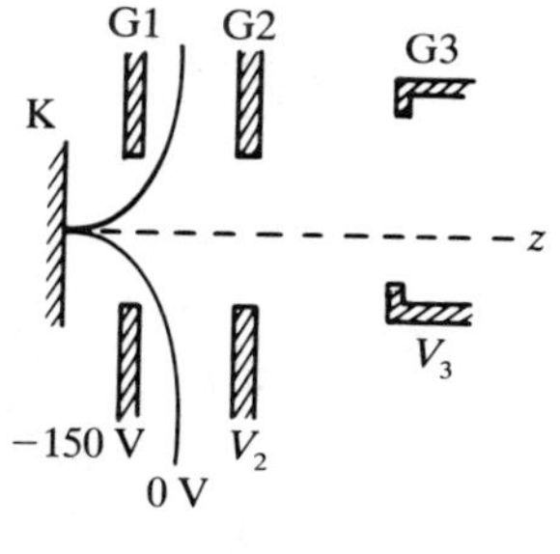

Fig. 11-17. The cut-off voltage situation when V_1 is set to its maximum value for fixed values of V_2 and V_3.

be reset at modified values in order to have cutoff at the same $V_{c0} =$ 150 V. In any event, V_2 and V_3 are at fixed values that depend upon the distance of these lenses from G1 and upon the radii and thicknesses of their apertures. Typical values for V_2 are in the 200–800 V range when V_3 is in the range of 5–10 kV and when the separation distance between these two lenses is 1–2 mm. These numbers are only approximate and are not meant to be taken as strict limits for cutoff voltage $V_{c0} = 150$ V.

Modulation of the current then occurs by varying V_1 between $-V_{c0}$ and some higher voltage, typically zero. When $V_1 > -V_{c0}$, the zero-volts equipotential shown in Fig. 11-17 will "open up" to resemble the one sketched in Fig. 11-16. The positive voltage difference $V_{c0} - |V_1| \equiv V_{dr}$ is known as the *drive voltage*. A beamforming region then has a specific *drive characteristic* relating current I to the drive voltage V_{dr}:

$$I = \kappa V_{dr}^{\gamma} \,. \tag{11.30a}$$

Some attempts have been made to derive γ from "first principles". Specifically, Gold and Schwartz (1958) give a formula for γ that appears to have the value $\gamma = 2.5$ as a lower bound, but it requires knowledge of the precise radial profile of the field $E_z(\rho, 0)$ close to the axis. Experience with very many BFR structures similar to that in Fig. 11-17 has yielded measurement values $2.5 < \gamma < 3.2$. Because the constant κ, different for different BFRs, is hard to determine, it is useful to express the quantities in Eq. (11.30a) with respect to reference current I_m resulting from a reference drive voltage (usually chosen at $V_1 = 0$, i.e., with $V_{dr} = V_{c0}$):

$$I = I_m \left(\frac{V_{dr}}{V_{c0}} \right)^{\gamma} \,. \tag{11.30b}$$

Parameter γ is actually not quite constant; it tends to change somewhat at higher currents from its almost constant value at lower currents, but the log-log plot of Fig. 11-18 is quite characteristic for the BFR of a kinescope electron gun. Equation (11.30) enable us to estimate the functional dependence of the cathode-emission radius $\rho_c(I)$ upon current I. Specifically, we obtain from Eq. (11.28) that

$$\rho_c \propto I^{1/2} [J(0)]^{-1/2} \,, \tag{11.31}$$

and by combining inputs obtained from Eqs. (10.52) and (10.73),

$$J(0) \propto V_A^{3/2} \,, \tag{11.32}$$

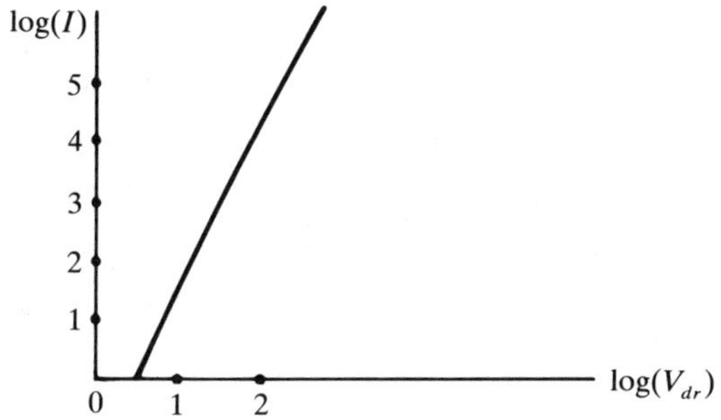

Fig. 11-18. The drive characteristic I vs V_{dr} (on a log-log scale).

where V_A is an *equivalent* anode voltage that we extrapolate from the situation immediately beyond the cathode shown in Fig. 11-16 (See section 11.6.2). Upon some reflection, one can conclude that V_A must be proportional to V_{dr} and therefore that Eqs. (11.31) and (11.32) must give

$$\rho_c \propto I^{1/2} V_{dr}^{-3/4} . \tag{11.33}$$

Combined with Eq. (11.30), this then yields

$$\rho_c = \rho_c^{(m)} [I/I_m]^{1/2-3/4\gamma} , \tag{11.34}$$

where $\rho_c^{(m)}$ is the reference radius for cathode emission at reference current I_m. Making use of the experimentally observed fact that $2.5 < \gamma < 3.0$ for a large range of current values and for a wide variety of BFRs, we observe that the dependence of $\rho_c(I)$ upon current is of the order of $I^{0.2}$ to $I^{0.25}$, which is what we made use of in the discussion of optimization in Section 11.5.4.

The above-described situation refers to modulation of current by *grid drive*. Modulation can also be effected by *cathode drive*, i.e., by maintaining the G1 grid at zero potential and decreasing the cathode potential V_c from the cutoff value $V_c = V_{c0}$ to a value that is lower. Equation (11.30a) will still hold with essentially the same κ if the grid geometry is unchanged, but now the drive voltage is $V_{dr} = V_{c0} - V_c$ for given $V_c \leqslant V_{c0}$. In order to understand cathode drive, let us examine a number of related situations summarized in Table 11-2 and discussed in more detail below.

In line 1 we have a situation of grid drive with $V_1 = 0$ so that $V_{dr} = V_{c0}$ in which case Eq. (11.30a) predicts the current $I_m = \kappa V_{c0}^{\gamma}$ is extracted from the cathode. In this case, the G1 grid plays little role other than physically blocking the path of some edge electrons emitted at unfavor-

TABLE 11-2

Line no.	K	G1	G2	Comment
1	0	0	V_2	Grid drive at reference $V_1 = 0$
2	0	$-V_{c0}$	V_2	Cutoff at grid drive
3	V_{c0}	0	$V_2 + V_{c0}$	Cutoff at cathode drive
4	0	0	$V_2 + V_{c0}$	Cathode drive at reference $V_c = 0$
5	V_1	0	$V_2 + V_{c0}$	Cathode drive at $V_c = V_1$

able angles with relatively high thermal velocities. Most of the beam will pass through the almost-zero curvature part of the V_2 equipotential plane closest to the axis. It can be seen from Figs. 11-16 and 11-17 that this part of the equipotential will lie closer to the cathode than the G2 grid itself. With the possible exception of very low-level currents, the emission between the cathode and this plane can be regarded as space-charge limited current in an equivalent planar diode with radial variation, e.g., as discussed in Section 10.5. Crossover of trajectories occurs close to the cathode and beyond this region. It is therefore plausible that the emitted current also obeys the Child-Langmuir law, $I_m = \kappa_1 V_2^{3/2}$, although the parameter κ is not known. Combining this with the above-cited prediction that $I_m = \kappa V_{c0}^{\gamma}$, we extract the relationship (also given by Moss 1968),

$$\kappa_1 = \kappa V_{c0}^{\gamma} V_2^{-3/2} \,. \qquad (11.35)$$

Line 2 summarizes the condition for emission cutoff at grid drive ($V_{dr} = 0$). Line 3 is equivalent to Line 2 because the electron optics force fields are unaffected by adding a constant potential V_{c0} to the cathode and to all grid voltages. It must therefore describe cutoff at cathode drive, and now we see that the G2 voltage should be maintained at $V_2 + V_{c0}$. Line 4 pertains to the reference situation for cathode drive ($V_c = 0$, $V_1 = 0$). By virtue of the arguments stated above, we find a concomitant reference current $I'_m = \kappa_1 (V_2 + V_{c0})^{3/2}$. If κ_1 is eliminated from this by applying Eq. (11.35), we then find

$$I'_m = \kappa V_{c0}^{\gamma} \left(\frac{V_2 + V_{c0}}{V_2} \right)^{3/2} = I_m \left(\frac{V_2 + V_{c0}}{V_2} \right)^{3/2} , \qquad (11.36)$$

in which the second form follows from Eq. (11.30a) with $V_{dr} = V_{c0}$. Finally, Line 5 shows the cathode driven at $V_{dr} = V_{c0} - V_1$, i.e., with voltage V_1 on the cathode. Because the geometry is still the same, Eq. (11.30a) still holds and the emitted current is $I' = \kappa V_{dr}^{\gamma}$ with essentially maximum (or reference) current when $V_c = 0$ or $V_{dr} = V_{c0}$ so that $I'_m = \kappa V_{c0}^{\gamma}$. Consequently,

$$I' = I'_m \left(\frac{V_{dr}}{V_{c0}} \right)^{\gamma} = \left(\frac{V_2 + V_{c0}}{V_2} \right)^{\gamma} I_m \left(\frac{V_{dr}}{V_{c0}} \right)^{\gamma} = \left(\frac{V_2 + V_{c0}}{V_2} \right)^{\gamma} I, \tag{11.37}$$

which demonstrates that the cathode-driven current differs from the grid-driven current by a constant factor $(V_2 + V_{c0})^{\gamma}/V_2^{\gamma}$. The relationship is borne out by a large body of BFR measurements (Moss 1968).

11.6.2 Estimate of Cathode Current

Our discussion of the behavior of electrons emitted from the cathode with a thermal distribution into the BFR should impress upon the reader that it is quite difficult to estimate the cathode current from the geometry and voltages of the BFR. There might be some possibility of achieving this feat analytically if only there were an anode-voltage surface that could be defined. However, it is discouraging that only the simplest diode configurations (which include planar, cylindrical, and spherical diodes) are amenable to rigorous analysis without recourse to brute-force numerical computation, whereas the BFR is a much more complicated structure. The possibility of an analytical treatment seems remote. To make matters even more complicated, the space-charge limited conditions that could lead to a Child-Langmuir law (or something related to that) hold only for a short distance beyond the cathode; for longer distance the electrons have been accelerated to velocities that decrease the electron densities to values too low for space-charge saturation. However, it is always possible to infer current from drive voltage by an experimental trial-and-error method of establishing a drive characteristic (I vs V_{dr}) with a reference drive voltage and current. The procedure is certainly not easy from an experimental point of view, and it is costly from a numerical point of view.

The purpose of this section is to suggest an alternative procedure that relies entirely upon Laplace (space-charge free) calculations, which are much less costly to do numerically. In fact, Laplace calculations are relatively easy to formulate and do from a numerical point of view because the significant part of the BFR volume in question is easily given Dirichlet conditions (values of the boundary potentials), except for gaps between grids, for application of a finite relaxation method (see Section 4.1.5). If the gaps between the grids intersect the boundary not too close to the apertures, then the potential drops linearly between adjacent grids so that Neumann conditions (specification of normal fields) can be given there. A full-scale Poisson calculation would be extremely difficult to carry out because the BFR has significant *longitudinal* space-charge forces, in contrast to the simplified transverse space-charge analysis of Section 10.3.

Let us consider again the situation sketched in Fig. 11-16, but under cathode-drive conditions. Thus, $V_1 = 0$, but the cathode voltage V_c is positive ($0 < V_c < V_{c0}$), and the equipotential labeled "0" is actually at V_c volts. Other potentials are all higher by V_c volts. Because space-charge limited conditions hold in the first few tenths of a millimeter beyond the cathode (perhaps only in the first 0.1 mm), and because most of the equipotential surfaces there are essentially planar at ρ values for which $J(\rho, z)$ is appreciable, the current will obey the laws derived in Section 10.5. We refer here specifically to Eq. (10.73) with J_s given by Eq. (10.52) in terms of an anode voltage V_0 and distance d which we will now rename as V_A and d_A, respectively, but which are not known.

We now describe an attempt at evaluating these quantities. The first step is to formulate a Laplace equation for the BFR situation defined by the possible existence of a beam during emission. Obviously, it is best done in cylindrical coordinates for axially-symmetric BFRs:

$$\frac{\partial^2 \Phi^{\mathrm{Lp}}}{\partial z^2} + \frac{1}{\rho}\frac{\partial}{\partial \rho}\left(\rho \frac{\partial \Phi^{\mathrm{Lp}}}{\partial \rho}\right) = 0 . \tag{11.38}$$

To solve the equation, we set $\Phi^{\mathrm{Lp}}(\rho, z) = V_c + V(z)W(\rho)$ and apply a separation-of-variables technique (see Section 4.1.2) to obtain for separation constant K^2:

$$\frac{1}{V}\frac{d^2V}{dz^2} - K^2 = 0 , \qquad \frac{1}{W}\frac{d}{d\rho}\left(\rho \frac{dW}{d\rho}\right) + K^2\rho = 0 . \tag{11.39}$$

The equation for $W(\rho)$ is Bessel's equation (Abramowitz and Stegun 1964), and it has only one finite solution at $\rho = 0$, namely $W(\rho) = J_0(K\rho)$, a Bessel function of the first kind of zero order. If this is to describe a situation as in Section 10.5 then we should choose $K = 2/\rho_c$ so that the zero equipotential occurs at $\rho \approx \rho_c$. Figure 11-16 indicates that this is a reasonable approximation (but "0" becomes "V_c" in cathode drive). The first equation in (11.39) has the general solution,

$$V(z) = v_1 \sinh\left(\frac{2z}{\rho_c}\right) + v_2 \cosh\left(\frac{2z}{\rho_c}\right) . \tag{11.40}$$

The constants v_1 and v_2 are easily obtained from the initial condition $V(0) = 0$ and from an imposed final condition that $V(d_A) = V_A - V_C$. When this is combined with the solution $W(\rho) = J_0(K\rho)$, we obtain

$$\Phi^{\mathrm{Lp}}(\rho, z) = V_c + (V_A - V_c)\,\frac{\sinh(2z/\rho_c)}{\sinh(2d_A/\rho_c)}\,J_0\left(\frac{2\rho}{\rho_c}\right). \tag{11.41}$$

The idea here is that d_A is an effective anode distance for the geometry imposed by the BFR with the hypothetical voltage V_A on the hypothetical effective anode (at least on the axis). The plots of equipotentials in the $\rho - z$ plane between the cathode and anode will resemble that of Fig. 11-16. To obtain a value of the anode distance d_A, as hypothesized above, for this configuration, we follow a suggestion first made by J.R. Fields (see de Wolf 1982), namely to define and evaluate the distance,

$$d = \left[\frac{\partial E_z^{\mathrm{Lp}}(0)}{\partial V_c}\right]^{-1}, \tag{11.42}$$

where $E_z^{\mathrm{Lp}}(0)$ is the longitudinal component of the Laplace electric field vector at the cathode (i.e., in the absence of space charge). This field can be obtained numerically, for example, by the method of finite-mesh relaxation (Section 4.1.5.1) for a BFR with grounded G1 ($V_1 = 0$) and fixed values of V_2 and V_3, as described above. By varying V_c (i.e., doing at least two Laplace calculations) an estimate of the d of Eq. (11.42) can be obtained. The next step is to take Eq. (11.41) and use it to determine the right-hand side of Eq. (11.42), namely,

$$E_z^{\mathrm{Lp}}(0) \equiv -\left[\frac{\partial \Phi(0, z)}{\partial z}\right]_{z=0} = \frac{2(V_c - V_A)/\rho_c}{\sinh(2d_A/\rho_c)}. \tag{11.43}$$

We take one more derivative of this to obtain the right-hand side of Eq. (11.42), and the result is,

$$d = \left[\frac{2/\rho_c}{\sinh(2d_A/\rho_c)}\right]^{-1}, \tag{11.44}$$

which is easily inverted to yield

$$d_A = \frac{1}{2}\,\rho_c \ln\left(\frac{2d}{\rho_c} + \sqrt{1 + \frac{4d^2}{\rho_c^2}}\right). \tag{11.45}$$

We now recall Eq. (10.73) for the space-charge limited current density in the *same* geometry, and we write it as

$$J(\rho) = \frac{2}{9\pi\beta} \frac{[\Phi(\rho, z) - V_c]^{3/2}}{[\bar{\rho}_c \sinh(z/\bar{\rho}_c]^2} , \tag{11.46}$$

noting specifically that $\Phi(\rho, z)$ has been replaced here by the difference $\Phi(\rho, z) - V_c$, also that $\bar{\rho}_c \equiv \frac{2}{3}\rho_c$, and that the $\sinh(d/\bar{\rho}_c)$ factor has been eliminated together with the $J_0(2\rho/\rho_c)$ factor of Eq. (10.73). If we apply this expression for $J(\rho)$ at the effective anode distance $z = d_A$, then $\Phi(\rho, d_A) = \Phi^{\mathrm{L}}{}_{\mathrm{p}}(\rho, d_A)$, and we can use Eqs. (11.41) and (11.43) to set

$$\begin{aligned} \Phi(\rho, d_A) &= V_c + (V_A - V_c)J_0\left(\frac{2\rho}{\rho_c}\right) \\ &= V_c = \frac{1}{2}\, \rho_c\left[-E_z^{\mathrm{Lp}}(\rho, 0)\right] \sinh\left(\frac{2d_A}{\rho_c}\right) . \end{aligned} \tag{11.47}$$

Finally, insert this into Eq. (11.46) with $z = d_A$ to obtain,

$$\begin{aligned} J(\rho) &= \frac{(2/9\pi\beta)[-E_z^{\mathrm{Lp}}(\rho, 0)]^{3/2}}{D^{1/2}} , \\ D &\equiv \frac{128}{81}\, \rho_c\, \frac{[\sinh(3d_A/2\rho_c)]^4}{[\sinh(2d_A/\rho_c)]^3} . \end{aligned} \tag{11.48}$$

Here then, we have an estimate of beam current obtained entirely from the space-charge free Laplace geometry. Note that

$$E_z^{\mathrm{Lp}}(\rho, 0) = E_z^{\mathrm{Lp}}(0, 0)J_0\left(\frac{2\rho}{\rho_c}\right) \approx E_z^{\mathrm{Lp}}(0, 0)\left[1 - \frac{\rho^2}{\rho_c^2}\right] . \tag{11.49}$$

Use this in Eq. (11.48) to obtain the total beam current,

$$I = 2\pi \int_0^{\rho_c} d\rho\, \rho J(\rho) = \frac{4}{45\beta}\, [-E_z^{\mathrm{Lp}}(0, 0)]^{3/2}\, \frac{\rho_c^2}{D^{1/2}} \tag{11.50}$$

It is interesting to compare this with Eq. (11.33) because here we observe that $-E_z^{\mathrm{Lp}}(0, 0)$ is proportional to V_{dr} and therefore current I is proportional to $V_{dr}^{3/2}\rho_c^2$, as we found in Eq. (11.32). This lends more credence to the arguments given in connection with Eqs. (11.31)–(11.33).

REFERENCES

Abramowitz, M. and I. A. Stegun (1964) *Handbook of Mathematical Functions*, National Bureau of Standards 55, U.S. Department of Commerce.

de Wolf, D. A. (1982) *RCA Rev.*, **43**, 339.

Glaser, W. (1952) *Grundlagen der Elektronenoptik*, Springer, Berlin.

Gold, R. D. and J. W. Schwartz (1958) *RCA Rev.*, **19**, 564.

Grivet, P. (1972) *Electron Optics*, 2nd edn., Pergamon, Oxford.

Harte, K. (1973) *J. Vac. Sci. Technol.* **10**, 1098.

Kirstein, P. T., G. S. Kino and W. E. Waters (1967) *Space Charge Flow*, McGraw-Hill, New York.

Kittel, C. (1966) *Introduction to Solid State Physics*, 3rd edn., John Wiley, New York.

Mayer, J. E. and M. G. Mayer (1966) *Statistical Mechanics*, John Wiley, New York.

Moss, H. (1968) Narrow angle electron guns and cathode ray tubes, in *Advances in Electronics and Electron Physics* (Supplement 3), Academic Press, New York.

Pierce, J. R. (1954) *Theory and Design of Electron Beams*, Van Nostrand, Princeton, NJ.

Scherzer, O. (1936) *Z. Phys.*, **101**, 23.

Washino, S., Y. Ueyama and S. Takenobu (1979) *IEEE Trans. Consumer Electron.*, **25**, 481.

Wilson, I. M. (1975) *IEEE Trans Consumer Electron.*, **21**, 32.

APPENDIX I

CONSERVATION OF THE 2-D LAPLACE EQUATION

The purpose of this appendix is to outline explicitly how the two-dimensional Laplace equation $\Delta\Phi = 0$ is conserved in a Schwarz-Christoffel transformation from $\zeta = x + iz$ to $w = u + iv$, as described further in Section 4.1.4. We could, of course, refer to the general properties of analytic functions, but we believe it is useful to spell it out somewhat here.

Let $w \equiv u + iv = F(\zeta)$ be the mapping of $\zeta \rightarrow w$. If $F(\zeta)$ is analytical, then $F'(\zeta)$ is uniquely defined as the derivative of $F(\zeta) = F_r(\zeta) + iF_i(\zeta)$ with respect to ζ. It is then easily seen that

$$du = F_r' \, dx - F_i' \, dz \, , \qquad dv = F_i' \, dx + F_r' \, dz \, . \tag{I.1}$$

This immediately leads to

$$\frac{\partial u}{\partial x} = \frac{\partial v}{\partial z} \, , \qquad \frac{\partial v}{\partial x} = -\frac{\partial u}{\partial z} \, . \tag{I.2}$$

As angles are preserved in a conformal transformation (see discussion in Section 4.1.4) we have $du \, dv = dx \, dz$ (the angle is 90° here), and from the Jacobian of (I.1) we must have $(F_r')^2 + (F_i')^2 = 1$. Consequently, $(\partial u/\partial x)^2 + (\partial v/\partial x)^2 = 1$. This property is crucial, together with the following,

$$\begin{aligned} \frac{\partial}{\partial x} &= \frac{\partial u}{\partial x}\frac{\partial}{\partial u} + \frac{\partial v}{\partial x}\frac{\partial}{\partial v}\,, \\ \frac{\partial}{\partial z} &= \frac{\partial u}{\partial z}\frac{\partial}{\partial u} + \frac{\partial v}{\partial z}\frac{\partial}{\partial v} = -\frac{\partial v}{\partial x}\frac{\partial}{\partial u} + \frac{\partial u}{\partial x}\frac{\partial}{\partial v}\,, \end{aligned} \tag{I.3}$$

in establishing that $(\partial^2/\partial x^2) + (\partial^2/\partial y^2) = (\partial^2/\partial u^2) + (\partial^2/\partial v^2)$, which completes the outline of the proof.

APPENDIX II

SOLUTION OF EQUATION (8.10)

Here we outline the procedure by which an approximate solution is obtained of the inhomogeneous second-order total differential equation,

$$\delta\rho'' + \frac{1}{2}\frac{\Phi_0'}{\Phi_0}\,\delta\rho' + \frac{1}{4}\frac{\Phi_0''}{\Phi_0}\,\delta\rho = \mathcal{R}_3(\rho_p, \rho_p'; \Phi_0)\,. \tag{II.1}$$

The approximate solution is obtained from known solutions of the homogeneous equation (which follows from the above by setting $\mathcal{R}_3 = 0$). Two useful independent solutions of the homogeneous equation are $\rho_1(z)$ and $\rho_2(z)$ with initial conditions,

$$\rho_1(z) = 1\,, \qquad \rho_1'(z) = 0\,, \qquad \rho_2(z) = 0\,, \qquad \rho_2'(z) = 1\,. \tag{II.2}$$

The technique for solving (II.1), using these two special solutions of the homogeneous equation, is *variation of parameters*. It consists of introducing two new functions $\alpha_1(z)$ and $\alpha_2(z)$, and assuming that the solution of (II.1) can be written as

$$\delta\rho(z) = \alpha_1(z)\rho_1(z) + \alpha_2(z)\rho_2(z)\,, \tag{II.3}$$

under the constraint that for all z

$$\alpha_1'(z)\rho_1(z) + \alpha_2'(z)\rho_2(z) = 0\,. \tag{II.4}$$

The solutions for the two new functions $\alpha_1(z)$ and $\alpha_2(z)$ then can be found by taking the derivative of both sides of (II.3) twice (and using the constraint (II.4)):

$$\delta\rho' = \alpha_1\rho_1' + \alpha_2\rho_2' ,$$
$$\delta\rho'' = \alpha_1\rho_1'' + \alpha_2\rho_2'' + \alpha_1'\rho_1' + \alpha_2'\rho_2' . \tag{II.5}$$

Insert these expressions into (II.1) and make use of the fact that $\rho_1(z)$ and $\rho_2(z)$ obey the homogeneous equation

$$\alpha_1'\rho_1' + \alpha_2'\rho_2' = \mathcal{R}_3(\rho_p, \rho_p'; \phi_0) ,$$
$$\alpha_1'\rho_1 + \alpha_2'\rho_2 = 0 . \tag{II.6}$$

These are two inhomogeneous linear equations in two unknowns, which are easily solved to obtain

$$\alpha_1'(z) = -\frac{\rho_2\mathcal{R}_3(\rho_p, \rho_p'; \Phi_0)}{\rho_1\rho_2' - \rho_1'\rho_2} ,$$
$$\alpha_2'(z) = \frac{\rho_1\mathcal{R}_3(\rho_p, \rho_p'; \Phi_0)}{\rho_1\rho_2' - \rho_1'\rho_2} . \tag{II.7}$$

The denominator of these expressions can be eliminated with the Sturm invariant (6.2) for ρ_1 and ρ_2:

$$\sqrt{\Phi_0(z)}[\rho_1(z)\rho_2'(z) - \rho_1'(z)\rho_2(z)] = \sqrt{\Phi_0(0)}\rho_1(0) , \tag{II.8}$$

so that we obtain Eqs. (8.13) from (II.7).

INDEX